FORD TUNING > SECRETS REVEALED

Written by experts in the field

Published by:

Kotzig Publishing

2505 NW Boca Raton Blvd

Suite 205

Boca Raton, FL 33431

send your comments to:

fordtuning@kotzigpublishing.com

All rights reserved. No part of this book may be reproduced or transmitted in any form or by any means, electronic or mechanical, including photocopying, recording or by any information storage and retrieval system, without written permission from the publisher, except for the inclusion of brief quotations in a review.

© Kotzig Publishing, Inc., 2001

Printed in Slovakia

ISBN 0-9715411-3-2

Disclaimer. This publication is being sold with no warranties of any kind, express or implied. This publication is intended for educational and informational purposes only, and should not be relied upon as a "how-to" guide in performing repairs, maintenance or modification of your vehicle. This publication is not endorsed by the manufacturer of your vehicle nor DiabloSport and there is no affiliation between the publisher and author and the manufacturer of your vehicle. Modifications performed on your vehicle may limit or void your rights under any warranty provided by the manufacturer of your vehicle, and neither the publisher nor the author of this publication assume any responsibility in such event. Any warranty not provided herein, and any remedy which, but for this provision, might arise by implication or operation of law, is hereby excluded and disclaimed. The implied warranties of merchantability and of fitness for any particular purpose herein are expressly disclaimed.

No Liability For Damages, Injuries Or Incidental, Special Or Consequential Damages. Under no circumstances shall the publisher, author, or contributing author of this publication, or any other related party, be liable to purchaser or any other person for any damage to your vehicle, loss of use of your vehicle, or for personal injuries suffered by any person, or for any incidental, special or consequential damages, whether arising out of negligence, breach of warranty, breach of contract, or otherwise.

State Law. Some states do not allow limitations of implied warranties, or the exclusion or limitation of incidental, special or consequential damages, so the above limitations may not apply to you. In such states, liability shall be limited to the greatest extent permitted by applicable law.

Warning. Although the publisher and author have made every effort to ensure accuracy of all information contained in this publication, the reader should not attempt any modifications of his or her vehicle or computer codes without the assistance of a qualified and trained technician. Always obey state and federal laws, follow manufacturers' operating instructions and observe safety precautions. All information contained in the publication is for educational and informational purposes only and is not intended as a "how-to" guide.

Note. All trademarks are property of their individual owners.

FORD TUNING

To my wife Allison,

for her love,

patience and support.

Contents

	Introduction	16
	Read a short description of each chapter	
Chapter 1 >	Dario Orlando get's his 380rwhp Mustang tuned (Contributed by Johan Mangs)	20
Chapter 2 >	How engines work (Contributed by Johan Mangs)	26
Chapter 3 >	Basics of Ford Electronic Engine Control (EEC)	36
Chapter 4 >	EEC program reader	50
Chapter 5 >	How the original EEC program was born	62
Chapter 6 >	Tuning software (Contributed by Patrick Stadjel)	72
Chapter 7 >	EEC program example (Contributed by Patrick Stadjel)	78
Chapter 8 >	Rom emulator	86
Chapter 9 >	Performance chip	94
Chapter 10 >	Handheld performance tool	104
Chapter 11 >	Measuring performance	112
Chapter 12 >	OBDII diagnostic (Contributed by Peter David)	116
Chapter 13 >	1500hp down the 1/4 mile with Willie Figueroa	142
Chapter 14 >	The PDQ story	148
	Cooking with DiabloSport, hot and spicy	152
	Appendix A: Dictionary of Automotive terms Locating a word in the book	166

ENTER

Preface

This book on Ford performance tuning secrets is for both expert tuners and absolute beginners I tried to write something for both worlds. This approach runs the risk of becoming too difficult for a beginner to understand and being too simplistic for an expert. My argument was, that no one knows everything; thus, reading a complete reference guide on the basics of tuning Ford vehicles would have something to teach everyone.

Let me tell you about my technical expertise. I have at least 50,000 hours of experience studying electronics and working as a chief engineer in the field of automotive diagnostic and performance. Numerous of my designs (or designs I participated in) are now in the hands of thousands of car repair technicians, making their lives easier. I learned, through my own personal experiences of taking my car to the service department (to ask what was wrong with my vehicle), how hard it is to get a accurate information. I have always felt compelled to share all the knowledge Ihave with others who might want to learn also, but in an easier manner. For future progress, any information collected should always be shared with others!

We should prepare ourselves for tremendous progress of electronics in all new automobiles. There is, currently, a lack of knowledge among technicians and enthusiasts. Car manufacturers are guarding all the software and hardware as intellectual property, releasing a bare minimum just to their own service departments.

Car repair has turned into parts swapping. Apparently, that is the way they want it. What does this mean for the automotive performance industry and the do-it-yourself enthusiasts? The newer the automotive technology, the harder time they will have understanding it.

That is why this book was born. I was inspired by a new Open Source movement in the computer software industry. I thought, why not do the same for the automotive industry. This book will fill a gap in your library, answer some of your questions about engine electronics and show you another path, allowing you to benefit from freedom of information.

While designing this book, I was pondering about the style of the finished product. I spent a lot of time in bookstores in the automotive section and never found a format I liked. I wanted a hard cover book, with quality paper and color throughout, that would lay flat and was made to last for years. I wanted both a How-to-do-it book and a resource guide.

I recalled from my university days how easy is it was to fall asleep when reading dry textbooks about the theory of electronics. How the electrical current flows from the positive pole to a negative, but wait, 30 pages later, is it really the other way around? Do the electrons flow from the negative pole to positive? Who cares? I wanted to read only about practical matters I could use in my day-to-day life. So that killed the idea of writing about tuning in a textbook format.

So, should it be made into an engineering databook? Well, I read many databooks and just recently studied everything about the OBDII diagnostic standard and I felt like I needed to hire a professor to translate it for me – it was beyond my Master's degree in Computer Science. Maybe it wasn't that bad, but I didn't enjoy reading it and it wasn't practical to use, either.

The style of the book "hit me" while cooking. One of my favorite cookbooks is The Regional Italian Cuisine by Reinhardt Hess. What a beauty! It is a combination of a How-to book and a guide to Italy. That makes it fun to read even when not cooking.

That settled it. Even if Ford Tuning Secrets Revealed will take much more effort, than it was originally planned, there will be no shortcuts! It will be informative, but practical.

It is not up to me to judge, whether I succeeded or failed. Since I wrote this book not for money, but for love of cars and electronics, I would love to hear from you. And, … thank you for buying this book.

Please, do me one favor - do not get it wet, while you bath, and do not lend it to anyone without writing his or her name down as a reminder.

If you enjoy reading it, you might want to monitor our website: *www.kotzigpublishing.com* for release dates of other books in the Secrets Revealed series:

- GM Tuning Secrets Revealed
  (July 2002)

- OBDII Diagnostic Secrets Revealed
  (July 2002)

Caution! Buckle up! The following pages reveal Ford Tuning Secrets!

Ivan J. Kotzig

ivanft@kotzigpublishing.com

Acknowledgements

This book, the first in
the Secrets Revealed series,
combines lifetime experiences
of various experts in the automotive
electronics performance industries.
They, too, learned something
during the process of creating this
guide; as no one can be an expert
in everything.

ENTER

ACKNOWLEDGEMENTS
FORD TUNING

Willie Figueroa, best known for his track record racing Ford Mustangs, is also a tuner at DiabloSport. He has been a World Ford Challenge Wild Street champion 3 years in a row (1999 - 2001); accompanied by many Pro 5.0, True Street and Mustang vs. Buick Grand National wins. Not only does he race the vehicle, Willie builds the drag car from the ground up, specializing in turbo/super-charging and nitrous oxide engine configurations. He contributed the chapter on measuring performance and solving common problems.

Patrick Stajdel, a tuner at DiabloSport, details about modifying the Ford stock program to create a custom one. His automotive experience started the day his father, a US Air Force colonel and pilot, had a vision of the future and obtained a chip programmer. He convinced Patrick to learn about weird things like hexadecimal numbers and lookup tables. Success came quickly - with his first modified chip he broke an IHRA national record in the 1986 Buick Grand National. After running his own speed shop and spending years working with three well-known names in the tuning industry - Hypertech, Superchips and currently DiabloSport - he is considered an expert on using tuning software and tuning emulators. He would like to thank his father for complete support and inspiration. (Patrick came up with the idea to print the "Confidential" watermark at the beginning of each chapter – we only did it to make him happy!)

SECRETS REVEALED

Peter David has been an automotive technician for 19 years, with expertise in European and Exotic vehicles. He has worked for fine dealerships such as Aston Martin, Jaguar and Mercedes-Benz for over 12 years, and spent an additional seven years as Shop Foreman for an independent repair facility. Three years ago, Peter joined the team at ToolRama, Inc. and has devoted his 19 years experience to the development of one of the leading diagnostic tools in the automotive world.

Johan Mangs, Operations Coordinator of DiabloSport, is an avid racing fanatic. Johan studied computer science and automotive technology. He is a performance chip and accompanying computer software specialist. To improve products and services, he gathers information from the field on a daily basis.

Who other, than Ruben Arnejo-Fernandez, the man behind the graphic image of DiabloSport, and our photographer, could produce the quality we needed? Ruben was born in Uruguay, and educated in classic drawing and painting; later studying both graphic and jewelry design. Early in his life he developed a fascination for owls and, thanks to a Gypsy priestess, avoided becoming an obsessed owl watcher.

(Ruben came up with the idea to put an owl on our Secrets Revealed stamp. He refused to remove it – so we left it there to make him happy!).

My mother, Margita Kotzig, my sister, Gabriela Kotzig-Erdos, and my nephew, Mathias Erdos, helped me tremendously with the recipes for our Cooking chapter. We know that you all will be as satisfied as the guests at our recipe-testing dinner party.

Direct any comments or complaints about the grammar in this book to Susan McCabe. Our VP and Chief Editor has over twenty years experience in the fields of publishing and advertising. On second thought…she may not take any complaints!

The other helpers whom I want to acknowledge are the management of DiabloSport, mostly the President Tawfik Zakak. They made every aspect of doing the research involved easy and helped to clarify the kind of book I wanted to write.

Introduction

Now you, too, can know the details of tuning Ford automobiles! This book is the first complete guide to electronic performance tuning of Ford vehicles both for absolute beginners and advanced tuners alike. If you want to understand the principles of operation of Ford Electronic Engine Control, this book is for you! Detailed descriptions will answer all of your questions and more.

It is the most comprehensive collection of information on Ford electronic tuning ever printed. Now the secrets, hidden in the machinery of only a few companies, can be yours!

Author Ivan J. Kotzig, longtime designer of automotive electronics, presents this interesting subject in layman's terms.

Ivan has a Master's degree in Engineering from the Hungarian Technical University and has dedicated his life to creating many inventive electronic circuits in the car repair industry.

Ivan is also a co-founder of several recognized firms in the automotive industry:

- ProgRama, a leading remanufacturer of automotive electronics

- ToolRama, a creator of handheld diagnostic tools

- DiabloSport, a manufacturer of performance electronics

ENTER

About this book

What is in this book

This book presents the principles of Ford electronic tuning. A book like this doesn't have to be read from cover-to-cover, regardless of the author's dream that you do so!

ENTER

1

This chapter shows you a picture-document of a real life tuning from begining to end. The story takes you to Florida, where the owner of Steeda Autosports, Dario Orlando, gets his street-fighting Mustang dyno-tuned by the DiabloSport crew.

Chapter 1 > Steeda's Dario Orlando gets his 380rwhp Mustang tuned

contributed by Johan Mangs

FORD TUNING

This chapter will show you tuning, step-by-step: the Ford Electronic Engine Control computer is removed from Dario's car; a newly made custom tune is programmed into a performance chip; the chip is installed into the computer and all the work is verified on a car dynamometer. If any of these steps are intriguing to you and you want to know about it in more detail, a text found next to each picture directs you to the appropriate chapter.

Steeda Autosports, located in sunny Pompano Beach, Florida, specializes in enhancing the performance of Ford Mustangs, trucks and Sport Utility Vehicles. Steeda spans the full spectrum of vehicle performance upgrades, from professional and amateur racing to new turnkey vehicle packages, retail and mail order part sales, plus design and manufacturing operations. Steeda was founded in 1988, when Dario Orlando saw an opportunity to put his (then) 15 years experience repairing and racing cars to work turning the Ford Mustang into a world class sports car. Twelve years later, Steeda stands at the pinnacle of the Ford performance aftermarket, with 34 full time employees and parts distribution to all 50 states through wholesale distributors.

Date: Tuesday, December 11th, 2001, 09:00am, 70Fahrenheit, 60% air humidity.

Location: Steeda Autosports, Pompano Beach, Florida.

The Vehicle: 2001 Ford Mustang GT, 4.6L SOHC 2-valve V-8, manual transmission, EEC-V with GTG3 code.

Equipment: Vortech S-trim centrifugal supercharger; 6-7psi of boost, Ford Racing Performance Parts (FRPP) #30 fuel injectors; 70mm throttle body; Steeda Ultra cool radiator; K&N air filter; Pro-M 80mm Mass air flow meter, calibrated for #30 injectors; Steeda Autosports high performance exhaust system, eliminating two of the four catalytic converters; 373 ring and pinion gears; Auburn differential; full Steeda G-trac suspension and brake upgrade; Steeda TRIAX shifter; Steeda ultra lite 17x9 aluminum wheels.

Owner: Dario Orlando, Steeda Autosports.

First the EEC was removed from the vehicle and the factory file was read out using an EEC reader. *(See chapter about EEC program reader).*

Once the vehicle has the proper equipment installed, it is time to fine-tune it for maximum performance and drivability. It is pointless to fine-tune a vehicle unless the right combination of equipment is installed and tested to work properly.

To monitor total performance gains, a dyno run was made while the vehicle was completely stock, referred to as the base run.

Mounting an air/fuel meter is necessary for monitoring the air/fuel ratios at all throttle positions during the dyno runs. The proper location is to utilize the factory O2 sensor location; acceptable, but not as accurate, is to mount it at the end of the tail pipe. The ideal air/fuel ratio for a supercharged Mustang of this caliber is 11:8 compared to 12:5-13.0 for non-supercharged, at wide-open throttle. This is a controversial subject and is very dependent upon vehicle, tuner preferences and fuel used (in our case 93 octane).

Securing vehicle on dyno utilizing secure mounting points to prevent the vehicle from moving around on the rollers during the dyno runs. Axles and frame serve as good secure mounting points. *(See chapter about Measuring performance).*

To calculate torque output, a sensor is clipped on to the tachometer signal wire, which feeds into the dyno interface and is then displayed on the dyno run readout.

Connecting ROM emulator to monitor and tuning EEC in real time. The ECC was removed and its contacts were properly cleaned in the same fashion as for a performance chip installation. *(See chapter about ROM emulator).*

The Hellion diagnostic scanner from DiabloSport was connected to the OBDII connector for monitoring the ignition advance in real time. *(See chapter about OBDII diagnostic).*

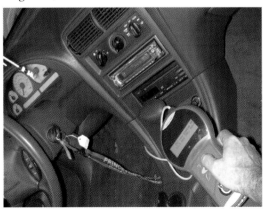

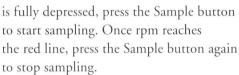

First run. Putting the vehicle in gear and working up to 4th gear, on 5-speed manual transmission, or 3rd gear with the overdrive off, on automatic transmission. As the throttle is fully depressed, press the Sample button to start sampling. Once rpm reaches the red line, press the Sample button again to stop sampling.

Analyze the dyno run chart for smoothness and peak horsepower and torque. Also reference the air/fuel ratios to the power curve.

If the peak hp or torque is not satisfactory, the tuner will need to change some parameters and repeat

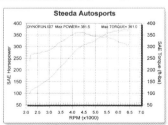

the previous two steps. *(See chapter about Tuning software).*

Once "safe" maximum peak hp and torque are achieved on the dyno, the vehicle is taken for a street test.

With the ROM emulator still connected, simulated every-day driving is monitored. Idle, speedometer errors, detonation and all other parameters are monitored carefully.

If the car feels right during the test drive, the tuner will save the current file on his laptop under a new name (in this case the factory file GTG3 will become GTG3SC73). It will later be programmed into a performance chip.

The new tune is programmed into a DiabloSport 4-bank chip. *(See chapter about Performance chip).*

This chapter is a basic overview of all components found under the hood of any modern Ford. It will turn your attention to the single component of importance for our purpose: the Ford Electronic Engine Control computer (EEC).

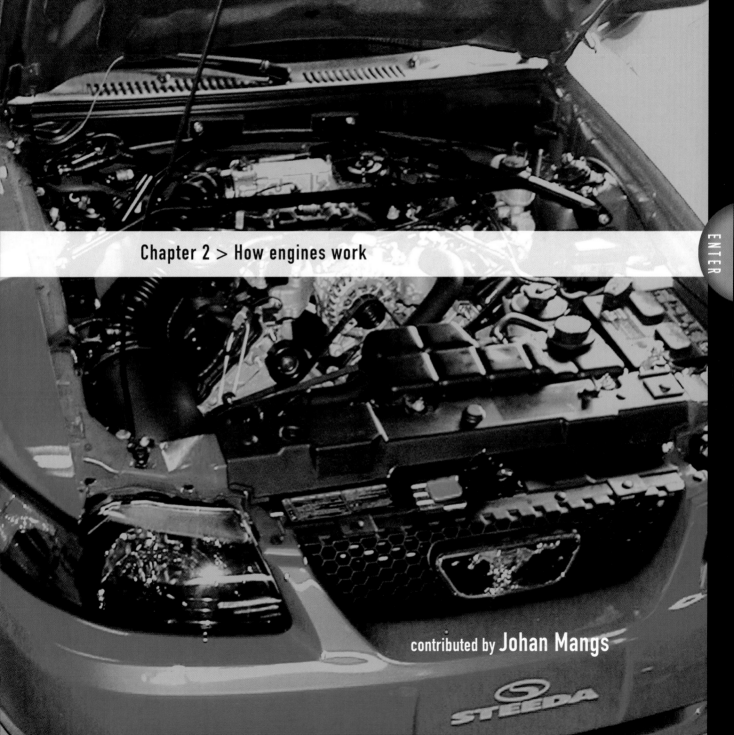

Chapter 2 > How engines work

contributed by **Johan Mangs**

Belgium is famous not only for it's chocolate in 1860 Belgian Etienne Lenoir invented the internal combustion engine, where the energy from an explosion was converted into rotary motion. The then used two-stroke combustion cycle was improved with the smoother running four-stroke cycle by Nikolaus Otto in 1876. With modification here and there, it is, even today, the most practical power source for automobiles.

The invention was soon used in real life. Before the end of the 19th century, Europeans were building very expensive handmade automobiles for sale to the public; only very few could afford them.

It was in America where it all changed. In 1908, Henry Ford launched production of the famous Model T Ford. When he introduced the moving assembly line, the car became much more affordable and soon half of all the cars in the world were Fords.

(Did you know? Henry Ford would only accept parts from vendors who placed them on pallets of specified size and wood, so he could reuse them for the floor boards of his Model T cars).

An internal combustion engine is basically an advanced air pump, i.e. the faster the air is allowed to pass through the engine the more power the engine will make. There are many factors that interact with each other to make this process complete and in order to gain maximum power (efficiency) every component must be synchronized to perform at exactly the right moment.

We focus on the four-stroke combustion cycle since there are no Ford vehicles with two-stroke engines produced today.

Four-stroke combustion cycle

The Intake Stroke: The piston starts at the top, also referred to as top dead center (TDC), the intake valve opens, and the piston moves down to let the engine take in a cylinder-full of air and gasoline. The amount of fuel mixed into the gasoline is determined by the amount of air the mass air meter reads.

The Compression Stroke: The piston then moves back up to compress this air/fuel mixture. By compressing the already volatile air/fuel

FOUR-STROKE COMBUSTION CYCLE

Induction stroke — **Compression stroke** — **Power stroke** — **Exhaust stroke**

mixture the volatility factor is increased tremendously making the combustion more powerful.

The Combustion Stroke: Right before the piston reaches the top of its stroke, the spark plug emits a spark to ignite the gasoline. The compressed air/fuel mixture in the cylinder combusts, gas expansion occurs, driving the piston down.

The Exhaust Stroke: Once the piston hits the bottom of its stroke, the exhaust valve opens and the exhaust is forced out to the tail pipe.

Chapter 2 > How engines work

FORD TUNING

Gasoline basics

Gasoline is often overlooked in high performance vehicles, the quality and octane rating of gasoline is extremely important, especially when running high performance engines.

Gasoline consists of a complex mixture of hydrocarbons and is most often produced by the fractional distillation of petroleum, also known as crude oil.

The compressed air/fuel mixture in the internal combustion engine tends to ignite prematurely rather than burning smoothly. This is due to the burning characteristics of gasoline. Low octane gas displays faster burning characteristics than high-octane gasoline.

When the air/fuel mixture burns too fast, or when it is ignited by the spark too soon, an engine knock is produced. A knock is a characteristic rattling or pinging sound in one or more cylinders. The sound is best described as someone hitting the engine with a hammer. The octane number of gasoline is a measure of its resistance to knock.

Fuel injected gasoline engine

Camshaft Controls the opening and closing of the valves.

Catalytic converter A special filter to absorb exhaust pollutants.

Crankshaft Converts the downward motion of the piston and rod into rotating motion.

Cylinder A combustion container of an engine.

SECRETS REVEALED

EGR Valve Reroutes a small amount of exhaust gasses back to intake manifold, reducing the amount of oxygen in air-fuel mixture, thus reducing the combustion temperature. That results in less oxides of nitrogen (NOx) in the exhaust gasses.

The EEC usually turns EGR off during idle or cold conditions.

Electronic Engine Control (EEC) The most important component for our purpose of electronic tuning, a computer brain which controls all the functions of the engine.

Exhaust Directs hot gases away from the vehicle.

Exhaust valve(s) Allows for exhaust gases to escape to the exhaust.

Fuel injector A solenoid, injecting a precisely controlled amount of fuel into the engine cylinder.

When current flows through the coil inside, a needle is pulled off it's seat allowing the fuel to spray out. The longer time the injector is energized, the richer the mixture.

Fuel pressure regulator Insures that the fuel in the fuel rail is at constant pressure.

Fuel pump Continuously circulates the fuel from the fuel tank into the fuel rail and through the fuel pressure regulator back into the fuel tank.

Fuel rail Feeds all the fuel injectors with fuel regulated at a constant pressure.

Idle control valve An air bypass valve, controlled by the EEC, which allows a precise amount of air to flow around the throttle plate, keeping the engine at the predefined idle rpm.

Intake air temperature sensor Measures the temperature of the air entering the engine, usually in the intake manifold or in the air cleaner. The air temperature affects the air/fuel ratios.

Cold air is denser and contains more oxygen; therefore it requires a richer mixture.

Intake manifold Distributes the air to the intake valves.

Intake valve(s) Allows the air/fuel mixture to enter into the combustion chamber.

Knock sensor If the ignition timing advance is larger than ideal, the engine starts to knock (a warning sign) and possible engine damage can occur. To achieve maximum performance, the Ford EEC keeps the ignition timing at the border of ideal value, and uses a sensor to obtain early warning of engine knocking. A knock sensor outputs a voltage when it detects engine knock. The EEC then slightly retards the timing to compensate for it.

Manifold pressure sensor (or Barometric Pressure or Air flow sensor or Air mass meter or Mass Air Meter)

Measures pressure in the manifold. Whichever sensor is used, its purpose is to measure the amount of air entering the engine, so the Ford EEC can continuously adjust the proper amount of injected fuel in order to create ideal combustion.

Oxygen sensor A sensor to measure the oxygen content of the exhaust gas for the purpose of lowest emission. The oxygen sensor outputs a voltage in proportion to the air/fuel ratio.

The EEC reads the oxygen sensor and then it decides if it should richen or lean out the mixture.

The oxygen sensor functions only when hot (more than 500F). While the engine is cold, the EEC uses preset fuel/air mixture ratios stored in so called look-up tables.

Oxygen sensor heater Electrical heating element in the oxygen sensor which drastically improve cold-start emission levels. Otherwise it would take a couple of minutes for the cold sensor to be heated up by the engine and start to generate a signal.

Piston Converts expanding gas (combustion) into downward motion.

Power Booster (option) An engine power booster is an air compressor - either turbocharger or supercharger - forcing more of the air/fuel mixture into the cylinders to achieve higher compression, thus generating more power.

Turbocharger Powered by exhaust gases through a turbine wheel.

Supercharger An air blower mechanically driven by the engine via a belt driven mechanism.

Intercooler Cools the intake charge from the compressor (compressed air gets hot).

Spark plug A component generating a spark to ignite the air/fuel mixture in the engine. The spark plug has two electrical points close to each other. When a high enough voltage signal is applied to the spark plug, the current will jump the gap between the two points, creating a spark.

The time and duration of the high voltage signal is precisely controlled by the ignition section of the Ford EEC.

Throttle position sensor The Throttle Position sensor (TP) monitors the mechanical position and motion of the throttle. It indicates how much power the driver is looking for by both how far and how fast it opens.

Throttle valve Controls the amount of air entering cylinders. The higher rpm and the load of the engine, the higher the amount of air required to run the engine.

Injection
Air-Fuel Mixture The ratio of air and fuel injected into a combustion chamber. (14.7 parts of air to 1 part of fuel is the ratio where every gasoline molecule is combusted).

Ideal mixture For normal driving, the ideal ratio is 14.7:1. For wide-open throttle, the ideal ratio is 12.5:1.

Lean mixture Insufficient amount of fuel, causing a power loss.

Rich mixture Excess amount of fuel, causing a power loss.

Ignition
Retarded spark timing Insufficient amount of ignition advance causing a power loss.

Advanced spark timing Excessive spark lead causing a power loss.

Engine speed (rpm) Rotational speed of the engine, measured in revolutions per minute of the crankshaft.

Diesel engine

(Please refer to the gasoline engine sensors and components, as the function is very similar, or the same, for the diesel engine components).

Unsatisfied with the inefficient gasoline engines of his time, **Rudolf Diesel** developed the concept for the diesel engine and acquired the German patent in 1892.

The diesel engine is an internal combustion engine with a four-stroke cycle very similar to the gasoline cycle. The only difference is that instead of the fuel being compressed with the air, air is added when the diesel fuel is compressed. Diesel fuel is ignited by hot compressed air, no spark plug is used.

The major differences between the gasoline and the diesel engine are:

- A gasoline engine intakes a mixture of air and fuel, compresses it and ignites the mixture with a spark. A diesel engine takes in just air, compresses it and then injects fuel into the compressed air. No spark plug is needed since the heat created when compressing the air lights the fuel spontaneously.

- A normal gasoline engine compresses the air/fuel mixture at a ratio of 8:1 to 12:1. A diesel engine compresses at a ratio of 14:1 to 25:1. The higher compression ratio of the diesel engine leads to better efficiency, but it also require a more robust engine construction to withstand the high pressures produced within the combustion chamber.

- Gasoline engines today generally use either *throttle body injection*, where the fuel is injected in the throttle body, or port fuel injection, in which the fuel is injected just prior to the intake stroke (outside the cylinder). Diesel engines inject the fuel directly into the combustion chamber, hence the name direct-injection.

Diesel fuel

Not only do they smell different, diesel and gasoline are very different. Diesel fuel is heavier and contains more oil. Diesel fuel evaporates more slowly than gasoline because it is heavier, it is constructed with more carbon atoms in longer chains than gasoline - its boiling point is actually higher than the boiling point of water. Diesel fuel requires less refining, which is why it is normally cheaper than gasoline.

Diesel engines generally get better mileage than equivalent gasoline engines. This is due to the fact that diesel fuel has a higher energy density than gasoline.

SECRETS REVEALED

This chapter should satisfy your curiosity for what is inside that computer box you just removed from your vehicle - it takes you into the layman's world of electronics and feeds you the healthy knowledge base needed for true understanding of electronic performance tuning.

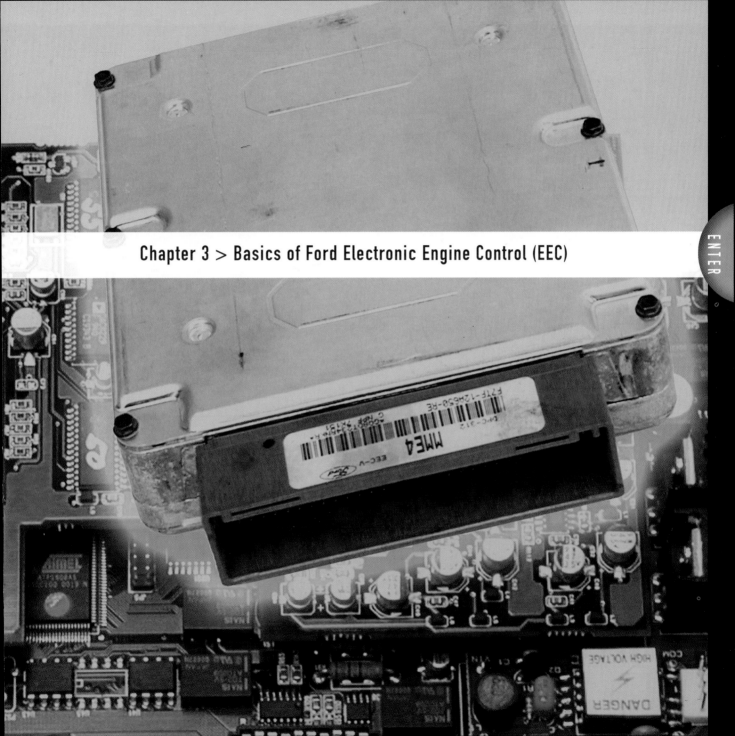

Chapter 3 > Basics of Ford Electronic Engine Control (EEC)

Circuit board and components

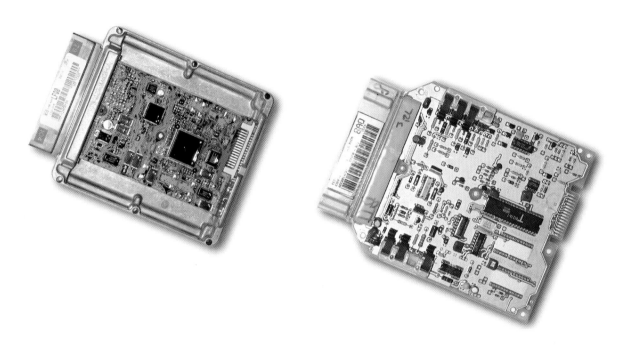

Harness connector

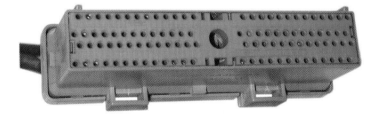

SECRETS REVEALED

Microprocessor

"A microprocessor is a Central Processing Unit (CPU) packaged in an Integrated Circuit (IC, or better known as a "chip")."

Let's dissect the above definition.

What is a Central Processing Unit?
It is the brain of any computer.

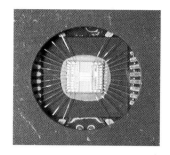

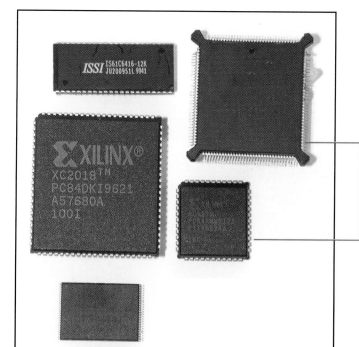

What is an Integrated Circuit?
Commonly known as a computer chip, it looks like a black plastic bug with a bunch of legs soldered into a circuit board. In reality the chip is only a tiny silicon piece enclosed within the IC itself. This silicon chip contains tens of thousands of microscopic transistors. Transistors are basic building blocks of any electronic circuit.

The legs of the bug, which are called pins, are internally attached to the silicon chip.

They protrude through the black plastic case in order to connect the silicon chip to the circuit board.

"The microprocessor executes instructions from a *program*, one at a time."

What is a program? A program is a procedure, written by humans, describing a task for a computer to execute. In our case, the task is to control the Ford engine.

What is a *human*? That is answered in some other publications.

In layman's terms, a microprocessor is the electronic brain of the Ford engine.

Read Only Memory (ROM)

"Read Only Memory (*ROM*) is an IC designed to permanently store the program and data for the microprocessor. This works like human long-term memory."

A more descriptive name for this kind of a memory would be "One Time Write, then Read Only Memory", but that wouldn't sound very good, would it?

There are various types of memories falling into this category. Whether they are called ROM, *PROM* or *EPROM*, makes no difference to us, unless we are the computer designer. The only thing of concern to us is that THIS is the chip containing the information we need to modify in order to achieve better performance from our Ford vehicle.

Similar to videotape, the information stored in ROM is *non-volatile*. That means it remains intact even if the power is off. Manufacturers of ROM claim a period of about 20 years of keeping data with the power off. Unlike the videotape, which wears down each time it is used, ROM gets refreshed for another 20 years each time it is powered up!

In layman's terms, the ROM contains the factory set engine control program. The focus of this book is to tune your program to achieve better performance.

Random Access Memory (RAM)

"Random Access Memory *(RAM)* is an IC designed to temporarily store data used by the microprocessor. This works like human short-term memory."

A more descriptive name for this kind of memory would be the original name "Read and Write Memory". This name ended up as unused as the metric system in the USA.

RAM will lose all data as soon as the ignition is turned off.

The Ford EEC also contains *KRAM* (Keeper RAM). It is an ordinary RAM, powered directly by the car battery. KRAM keeps the data even if the ignition is turned off.

Both RAM and KRAM will lose data if the car battery dies or is removed.

In layman's terms, performance tuners need not worry about altering data in these chips!

Bus

Bus is another name for multiple signal lines dedicated to the same task. The bus is like an electrical cable. The purpose of the bus is to connect one chip to another. In real life, instead of wires, etched copper connections are used.

The count of lines within a bus is called *Bus Width*. Bus width is measured in *bits*. The typical bus is 8, 16 or 32-bits wide. This means that if a bus is 16-bits wide, it has 16 etched lines on the circuit board.

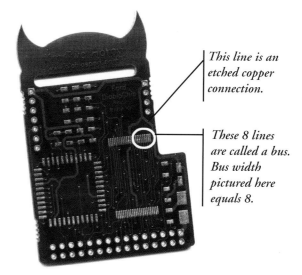

This line is an etched copper connection.

These 8 lines are called a bus. Bus width pictured here equals 8.

The bus connects the EEC's main chip (microprocessor), to other chips, like ROM and RAM, allowing them to share information.

The Integrated Circuit (IC, or simply a chip) sending the information through the bus is called a transmitter. Engineers call it "chip writes to the bus". Likewise, the IC receiving the information from the bus is a receiver, or simply "reads the bus".

Similar to traffic on the road, which can be one-way or two-way; data can flow through the bus one-way or two-way. The bus is like the road and data is like the traffic. Engineers like to call the two-way bus "bi-directional".

A computer needs more than just one bus. To distinguish one from another, buses are named by the signal they carry. Two very common buses are an *Address Bus* and a *Data Bus*.

In layman's terms, the bus is a bunch of wires linking one chip to another.

Purpose of the bus

Memory chips are able to store thousands of data pieces, called *bytes*. An example of a byte is one character. The English word "Hello" consist of 5 characters; therefore its size is 5 bytes.

Engineers call the whole content of the memory *data*. The individual pieces of the data are called *data bytes*. Each data byte is stored on a different location of the *memory chip*. These locations are called *addresses*.

In order to retrieve a byte from the memory, the microprocessor must first point to its location. The address bus is used to point to the location. The address bus is a one-way bus and only the microprocessor can write to it. One way to visualize it is that the address bus is a one-way road leading from the microprocessor.

The data bus is used to retrieve the data from the memory. The data travels from the memory to the microprocessor, the data bus can also be used to write new data to memory, in which case the data travels the other way – from the microprocessor to the memory. That means that the data bus is a bi-directional bus.

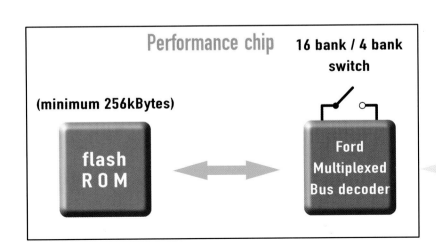
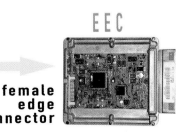

SECRETS REVEALED

 In layman's terms, the purpose of the bus is to connect multiple chips to allow them to talk to each other.

Multiplexed bus

In automotive computers in general, the larger the size of the EEC memory, the larger the address bus width.

 In layman's terms, a large EEC memory requires a large amount of connecting wires.

In the case of all Ford EEC computers, the memory is larger than the amount of available connecting wires. A secret: the EEC memory needs 24 wires, but EEC has only 8 available.

Using a *multiplexed bus*, Ford and Intel solved this problem. They used multiple bus-writes for the same single address. That saves on wire count, but slows down the process.

For example, if a 24-bit wide message is needed, but only an 8-bit wide bus is available, the microprocessor needs to write 3 times to the bus to send the entire message.

Ford multiplexed bus requires the use of a proprietary microprocessor, ROM and RAM. As a result, without having a custom-built chip reader, there is no way to read the program out of Ford the EEC ROM.

 In layman's terms, the Ford EEC ROM chip, which needs to be tuned, is non-standard and is not available to the general public. That is why the after-market developed a replacement; called a performance chip.

Ford Multiplexed Bus

Layman, if you don't care about the intricacies of the Ford Multiplexed Bus and the performance chip, you might want to skip this section.

We know that the Ford Multiplexed Bus is only 8-bits wide. Engineers are notorious for naming bus signals from the right to the left, so the signals are named MB7 through MB0, or simpler to write, MB [7..0]. *MB* stands for Multiplexed Bus.

In EEC-IV (vehicles from 1986-95) the capacity of ROM is 64KB. A microprocessor in a non-multiplexed system would need a16-bit wide address bus and an 8-bit wide data bus to be able to address any location in the same ROM. That adds up to a 24-bit wide information, or, simply, 24 wires. The Ford Multiplexed Bus has to take care of this by cutting it into three chunks of 8 bits each. The problem lies in distinguishing which part of the message is currently on the bus. Is it the first or the second part of the address, or perhaps the data itself? The Ford microprocessor indicates this with three additional output signals – *DIR, INSTR* and *STROBE*. Because the bus couldn't work without these, they are considered an integral part of the Ford Multiplexed Bus.

DIR stands for Direction. If the microprocessor wants to read current data from the bus, it will set DIR to logical 1.
If the microprocessor wants to write data to the bus, it will set DIR to logical 0. Other microprocessors on the market usually call this signal R/W, or Read/Write.

INSTR stands for Instruction.
If the microprocessor wants to fetch the next program instruction from the ROM, it will set INSTR to logical 1. If the microprocessor wants to read engine calibration data from the ROM, it will set INSTR to logical 0.

/STROBE stands for Strobe. Strobe is the heartbeat of the bus. At the very moment of a pulse on the /STROBE line, depending on the current value of DIR signal, one of two things is supposed to happen. If DIR=0 (meaning the microprocessor is writing the data), the memory chip is supposed to latch the 8 bits of information on the multiplexed bus. If DIR=1 (the microprocessor is reading the data), the microprocessor will latch the bus information loaded there by the memory chip.

In digital systems, it is customary to express the default (or idle) logical value of any signal by the way its name is spelled. If the default level is logical 0, the signal is called Active 1 (meaning it is active when it has a value of logical 1). There is no special mark in the spelling of their name. Both DIR and INSTR signals are active 1.

/STROBE signal is logical 1 by default (it has a value of logical 1 whether the signal is idle, or passive). That means /STROBE is Active 0 (strobe pulse is present when the signal level is logical 0). To indicate it in the spelling of its name, a '/' character precedes it.

In EEC-IV, the eleven signals of Ford Multiplexed Bus (MB[7..0], DIR, INSTR and /STROBE) will allow the microprocessor to access the entire ROM memory.
With the introduction of EEC-V, ROM capacity was quadrupled to 256KB.
The microprocessor on a non-multiplexed bus would need an18-bit wide address bus, which is 2 bits more than the Ford Multiplexed Bus can offer. To solve this, engineers at Ford and Intel added two more signals, *BS0* and *BS1*.

BS stands for Bank Select. Signals BS0 and BS1 can be set to four different logical combinations (00, 01, 10 and 11), allowing the microprocessor to select which quarter (or bank) of the ROM will be accessed.

Analog and digital circuits

In pre-EEC times there was no need for any bus. That was the Analog Age. One signal wire would carry a huge amount of combinations of data. Analog data changes amplitude over time. Measuring it with a voltmeter, it can have any value, in the case of automobiles, it is usually between 0 and 14.4 Volts (that is going to change very soon – your next Ford will have a 36 Volt battery!). An example of an analog line is an audio output wire from the radio receiver to the speaker.

Today we live in the Digital Age. Digital data lines can, unfortunately, carry only one of two possible values. Measuring it with a voltmeter, in a 5 Volt digital system, it can only have a value of 0 or 5 Volts. The actual values might differ slightly, but the digital circuit will ignore the variation and round it off to the nearest value. If the voltage is below 2.5 Volts, it will treat it as a logical 0, or else it will be treated as a value of logical 1. It is thanks to this ignorance that digital circuits are so resistant to noise. (Almost every electronic gadget around you operates digitally).

The Ford EEC contains both analog and digital circuits. Most wires coming out of the EEC are connected to analog sensors (water temperature, mass-air meter, oxygen sensor, etc.) and analog electrical components (fuel pump, fuel injector, etc.). Altering any of these external devices, without also modifying the EEC program, will cause the engine to run poorly (higher exhaust emissions, trigger the Check Engine light) and will usually result in decreased performance.

In layman's terms, when modifying the EEC program, we are tuning the digital section of the EEC. When modifying any external components (mass-air meter, fuel injector, etc.), we are tuning the analog section of the EEC. Modifying both at the same time yields the highest level of performance.

EEC types

8061, a modified version of an off-the-shelf Intel microprocessor 8096, has a different pinout, a few more instructions, and a customized bus multiplexing. The addressing range (and therefore the program size limit) of 8061 is 64 KB.

ROM memory is EPROM type, in DIP package, labeled 8763. Capacity of ROM is 64KB (that is 64 kilobytes, which is about 64,000 bytes, but more about it is in the chapter explaining the original EEC program).

RAM memory, in a 24-pin DIP package, is labeled 81C61.

KRAM memory, in a 24-pin DIP package, is labeled 87C61.

EEC-IV
The custom microprocessor, in a 40-pin rectangular DIP package (with pins on two sides of the chip), is labeled 8061. Oscillation frequency is 15 MHz.

EEC-V

The microprocessor, in a 68-pin square PLCC package (with pins on all four sides of the chip), is labeled 8065. Oscillation frequency was increased to 18 MHz. The addressing range (and therefore the program size limit) was quadrupled to 256 KB, by adding two bank select pins, BS0 and BS1.

ROM type is *flash* ROM, in PLCC package, labeled either 81C62 or 81C65. Capacity of ROM is 128KB or 256KB.

J3 test port

The *J3 test port* is an interface to the Ford Multiplexed Bus.

Both the EEC-IV and the EEC-V are enclosed in an aluminum enclosure. On the opposite side of the harness connector is a small opening, covered by a visible sheet of metal or plastic insert. Visible behind this part of the EEC circuit board, with 15-pin contact pads on each side, usually called the J3 edge connector, the J3 service port or the J3 test port.

The J3 test port was not intended for aftermarket use, as it is covered with epoxy solder mask and grease. It was probably sealed in the factory by the quality control department after the EEC passed the final inspection.

Yet the J3 port offers the most convenient access to the Ford Multiplexed Bus and some other signals. It is commonly used by the aftermarket industry. **Before it can be used, it must be carefully cleaned.** The proven way of doing this is described in the chapter about installing the performance module.

Connector diagram

Pin numbering looking from the back of the EEC into the opening:

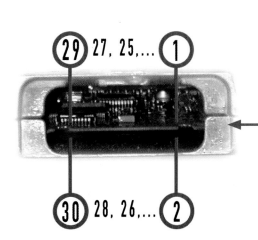

J3 Port definitions	
29 GND	30 GND
27 POWER (+12V)	28 POWER (+12V)
25 DIR	26 BS0
23 INSTR	24 BS1
21 /STROBE	22 PROGRAM
19 MB7	20 /RAM_disable
17 MB6	18 /Internal_ROM_disable
15 MB5	16 Erase
13 MB4	14 /Memory_reset
11 MB3	12 /Reset
9 MB2	10 /Pause_microprocessor
7 MB1	8
5 MB0	6
3 V5N	4 (high for access)
1 UREF (5+)	2 VCC (+5V)

Some versions of the EEC do not have all the signals routed to the J3 test port. Before tuning, small jumpers have to be soldered onto the circuit board:

After soldering

SECRETS REVEALED

Now that your Ford's EEC
is on your desk, you want
to read the factory
program from it.
If you care to know how
the equipment used
for this purpose works,
please read
this chapter.

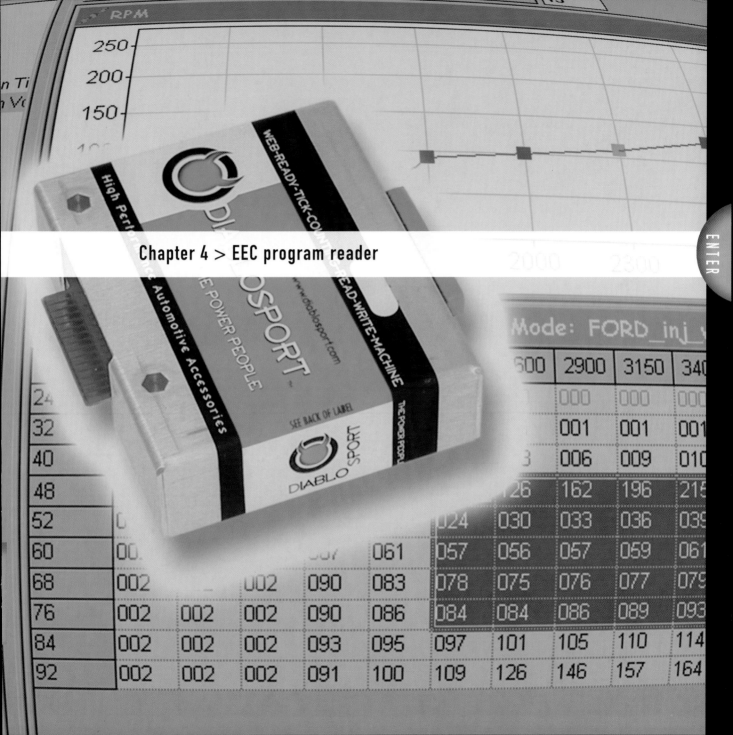

Chapter 4 > EEC program reader

Hardware

The process of electronic tuning requires a modification of the factory engine calibration data contained in EEC ROM.

The modification, or editing of data, is done on a PC. Regardless of which computer is used (desktop computer or a laptop), what operating system (DOS or Windows) and what editing software (binary editor or custom program) – it all begins with the original factory file.

The most convenient and accurate way of obtaining a factory file is to read it out of the EEC being tuned, using an EEC program reader.

An EEC Program reader is an interface between a PC and the EEC. An interface is necessary, since the Ford EEC doesn't have any other way to connect to a PC.

On the EEC side, the simplest way to connect is through the J3 connector. The other choice is to connect to the OBDII diagnostic plug, located inside the vehicle, somewhere within reach from the driver's seat (to read out the flash ROM through the diagnostic plug is somewhat more complicated and is described in the chapter about a handheld performance tool).

On the PC side, there is a choice of COM port (serial), LPT port (parallel) or USB (Universal Serial Bus). Each of the PC connections has its advantages and disadvantages, and it is up to the interface designer to select one. Once it is selected, custom PC reader software and custom interface *firmware* software must be written.

In layman's terms, the EEC program reader is a box, which connects the PC to the EEC. It allows us to read the factory EEC program.

How the EEC program reader works

The program reader interface can be built quite simply with a few logical integrated circuits. One could use widely available TTL 74xx type and avoid the need to write the firmware (software written for small microprocessor based microcomputers) for the microprocessor inside the interface box. Such a reader is even described on the Internet. A more flexible solution, described here, is to build a simple microcomputer.

The task of the interface is to receive a request from the PC to read the EEC data, and convert it to Ford Multiplexed Bus compatible signals. The reader interface reads data from the EEC, byte-by-byte, and continuously transmits it back to PC.

One must read 64KB of data from the EEC-IV, and 256KB of data from the EEC-V. There could be a hardware switch on the reader interface, or a selection on the PC software to select the read size. The user of such a reader would have to be careful to not to make a mistake and select the EEC-IV and actually read an EEC-V – only a quarter of the data would be read.

On the EEC side of the reader interface is a 30-pin female edge connector, mating with the J3 connector on the EEC. In order to be read, the EEC must be powered with 12 Volts. It is likely that the reader requires power as well. To eliminate a need for two separate power supplies, usually only the reader is powered and supplies the EEC with power through the J3 connector.

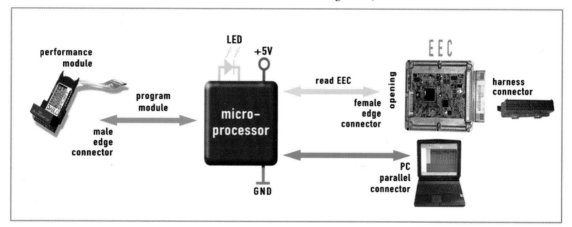

Another good option is to ignore the real size and always read 256KB of data. The advantage of eliminating the risk of user error outweighs the disadvantage of having unnecessarily larger data files – today's PCs have plenty of storage capacity. When 256KB of data is attempted to be read out of an EEC-IV, the true 64KB data is actually read four times and stored in one file.

A very important step, before the reader interface is connected to EEC, is to clean the EEC's J3 edge connector thoroughly. This is well illustrated in the chapter about performance modules. If the connector cleaning is not done properly and fragments of grease or epoxy based solder mask remains on the EEC's edge connector, connecting it to the reader will contaminate the reader's connector, rendering it useless. It could get so ugly, that the reader would need the connector to be exchanged – a task probably requiring the reader to be returned to the manufacturer, and most likely, not covered by warranty.

Beside the possible damage to the reader, an improperly cleaned EEC connector will not make good electrical contact with the reader and will prevent it from obtaining a useful read.

Not all of 30 pins of the J3 connector are used for reading. Obviously, to power the EEC, Power and Ground connections are necessary. To be able to read the ROM, all multiplexed bus connections (DIR, INSTR, /STROBE, BS0, BS1 and MB[7..0]) are needed. To be able to reset the EEC to a known state, /Memory_reset and /Reset connections are used. Finally, to be able to put the EEC microprocessor to sleep, /Pause_microprocessor connection is needed - otherwise the EEC microprocessor would interfere with the reading process.

The principle of actual reading is very simple. The reader, connected to internal the EEC multiplexed bus through the J3 connector, is practically replacing the internal EEC microprocessor. By applying a logical 0 to /Pause_microprocessor pin on the J3 connector, the EEC microprocessor stops its operation. The reader can now pulse both /Memory_reset and /Reset pins to clear the EEC ROM memory address counter.

In layman's terms, external equipment called the EEC Program Reader must be connected between the EEC and the PC. By manipulating the signal lines on the J3 connector, the reader takes control of all the chips in the EEC and reads out the contents of the EEC's ROM.

Reading the EEC

There are four possible banks to read from, selected by BS0 and BS1 pins. It doesn't matter in which order the banks are read out. It is purely a choice of the reader designer. Since engineers like to have an order in their creations, the designer will probably start by setting both lines to logical 0.

At this time the reader should apply power to the EEC. The reader interface should have a way to control the power to the EEC. It is good practice not to connect anything to the EEC J3 connector with the EEC under power – damage could occur.

The reader is initialized and ROM is ready to be read out at this point - a reading loop should begin.

The first desired address is written to the bus. Because addresses are 16-bits wide and the bus is only 8-bits wide, the address must be written to the bus twice, 8 bits at a time, using one strobe pulse for each write (remember – this is a multiplexed bus). The EEC ROM will receive the two 8 bit chunks and assemble them back to a complete 16-bit address. ROM will immediately post the requested data byte on the multiplexed bus. The reader will grab it and transmit it to the PC. This byte reading procedure can be repeated any amount of times. All the reader has to do is to increment the desired address by one and execute the loop again.

A standard 64KB memory address range is from 0000h to FFFFh (h stands for *hexadecimal* - hexadecimal numbers are explained in the chapter about understanding the original EEC program). The EEC ROM never contains data in the entire range. Reading all of the addresses would be a waste. For the EEC ROM desired addresses range from 2000h to FFFFh, which amounts to only 56KB.

Once the reader read out the last byte of current bank (address FFFFh), BS1 and BS0 bank select pins must by switched. If they were of logical value 00 for the first read, one could continue in order of 01, 10 and 11 to obtain data from all four banks. Next, the desired address is reset back to 2000h and the byte reading starts over.

After all 4 banks are read, reader interface logic should disconnect the power from the EEC and the PC software will typically notify the user of completion.

Four times the range of 2000h to FFFFh is 224KB (for those really precise it is actually 224 x 1,024 = 229,376 bytes). Each EEC read will have the same file size.

*In layman's term*s, the data in the EEC is organized in a somewhat complicated fashion and its size can vary. The most practical way to read any EEC program is to ignore its true size, assume it has the maximum size known for EEC (229,376 bytes) and simply read out all the data (it might be overkill, but who cares?).

DiabloSport reader / writer

One reader on the market is designed and manufactured by DiabloSport. It doubles as a performance module writer (or, more properly, programmer). Performance modules are described in their own chapter later in this book. For now, we can reveal this much;

after the original program is tuned on the PC, it must be written to a performance module by using a writer.

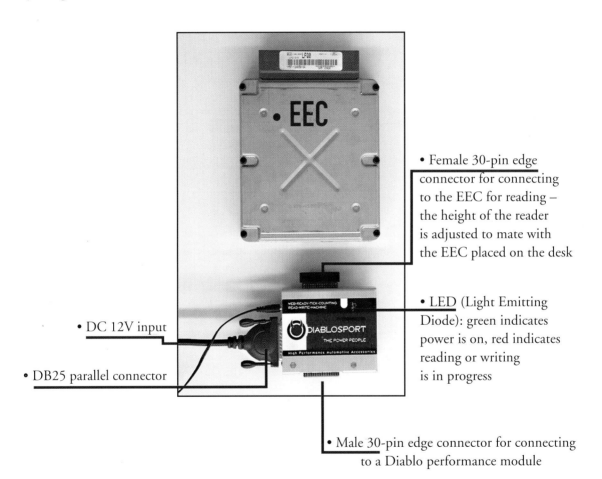

- Female 30-pin edge connector for connecting to the EEC for reading – the height of the reader is adjusted to mate with the EEC placed on the desk

- LED (Light Emitting Diode): green indicates power is on, red indicates reading or writing is in progress

- DC 12V input

- DB25 parallel connector

- Male 30-pin edge connector for connecting to a Diablo performance module

Reader / writer hardware

Layman, you might want to skip this chapter. It explains the detailed intricacies of the DiabloSport reader/ writer.

The DiabloSport reader / writer is powered by 12V / 1A wall-plug type transformer and has ample power to supply both its own circuit and the EEC when reading.

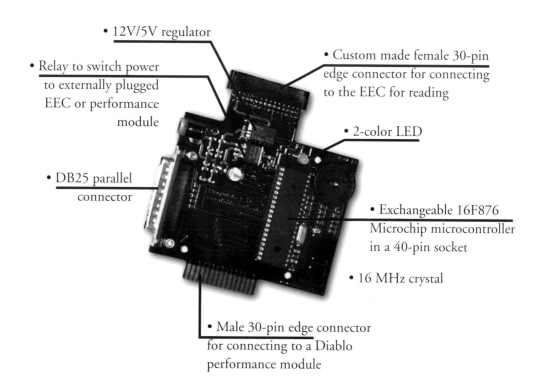

- 12V/5V regulator
- Relay to switch power to externally plugged EEC or performance module
- Custom made female 30-pin edge connector for connecting to the EEC for reading
- DB25 parallel connector
- 2-color LED
- Exchangeable 16F876 Microchip microcontroller in a 40-pin socket
- 16 MHz crystal
- Male 30-pin edge connector for connecting to a Diablo performance module

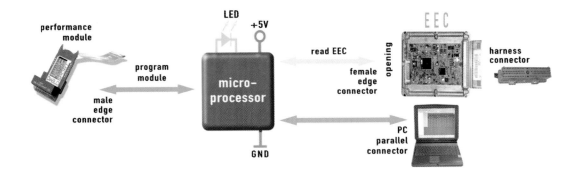

The heart of the circuit is a single-chip microcontroller made by Microchip. Detailed specifications are available to download from Microchip's website at www.microchip.com.

The reader / writer is powered by an external 12V transformer. The same power is routed through relay contacts to a female 30-pin edge connector to supply the EEC with power while reading.

Internally, the microcontroller requires 5V power. It is generated by the on-board 7805 linear regulator. The 7805 regulator has only three pins: input, ground and output. It has built-in short-circuit and thermal-overheat protection. Capacitors connected to both input and output of the regulator (are) supposed to filter out all the voltage surges and noise coming in from the external power

transformer. The 5V power generated by the regulator is also routed through relay contacts to a male 30-pin edge connector to supply the performance module with power while writing.

The microcontroller contains a program, called firmware, programmed into it at the DiabloSport factory. As soon as the reader is powered on, the microcontroller executes the firmware instructions at the rate controlled by the on-board 16-MHz (16,000,000 oscillations / second) crystal. The crystal oscillations are cleaned-up by the circuitry within the microcontroller and further divided by four, resulting in the final speed of executing 4,000,000 instructions / second.

After the initial power-up procedure, the microcontroller will turn on the green LED (Light Emitting Diode). The relay is still off and there is no power in either of the two edge connectors. The microcontroller is in the idling state.

While in the idling state, the microcontroller keeps monitoring the parallel port, which connects the reader, through standard DB25 parallel cable, to the parallel port of the PC. If any instruction is received from the PC, the microcontroller will exit the idle loop, decode the instruction and respond to it.

Reader / writer firmware

The simplest instruction is `get_reader_version`, issued by the PC as a request for reader type and version. The PC software, which is controlling all the actions of the reader, is continuously being improved. Registered owners of the reader can download it from the DiabloSport website at www.diablosport.com. Some PC software upgrades might require an upgrade of the reader hardware as well. Typically, that involves opening the reader, removing the old microcontroller from its socket and replacing it with a new one. Before the PC commands any work to the reader, it will issue a `get_reader_version` instruction to find out if the proper reader is attached.

If the PC approves the reader version, it will issue a `turn_power_ON` instruction; which, not surprisingly, will instruct the microcontroller to turn the relay in the reader on. At the same time, the reader will change the color of the LED from green to red, indicating to the user that reading or writing is in progress. Since power is now applied to both edge connectors, it is recommended that NOTHING is plugged or unplugged to or from the reader. A simple rule: do not touch anything until the light turns green – or something might get destroyed.

This brings up one point: the reader is not smart enough to recognize the mistake if both the EEC and the performance module are plugged in! Just live with it and only plug in one thing at a time, please.

After the reading or writing is completed, the PC must issue a `turn_power_OFF` instruction; which will instruct the microcontroller to turn the relay in the reader off. At the same time, the reader will change the color of the LED from red back to green, indicating to the user that the job is done.

Reading the EEC

To read a byte from an EEC: `set_EEC_to_ADDR`, followed by `read_EEC_byte` instructions are used.

The first instruction, `set_EEC_to_ADDR` (where ADDR is a variable), will instruct the microcontroller in the reader to manipulate all the necessary lines of the EEC multiplexed bus to set an address.

The second instruction, `read_EEC_byte` will instruct the microcontroller in the reader to manipulate all the necessary lines of the EEC multiplexed bus to read a byte from the EEC. Multiplexed addressing was described in the previous section of this chapter.

After a byte is read from the EEC, it is transmitted to the PC. This will repeat until all 4 banks are read. The DiabloSport reader always reads all 4 banks, which means it will read 229,376 bytes.

Write data to performance module

Data is written in a loop until all the bytes are written. A `set_module_to_ADDR`, followed by `write_module_byte` instructions are used.

The first instruction, `set_module_to_ADDR` (where ADDR is a variable), will instruct the microcontroller to set an address in the module.

The second instruction, `write_module_byte` will instruct the microcontroller to program a byte into the flash memory of the module. This is a well-known procedure, available to download from the Internet. There are many manufacturers of compatible flash memories and the procedure is basically always the same. You can browse the web and download the programming specifications from:

- AMD (www.amd.com/us-en),
- Atmel (www.atmel.com),
- ST Microelectronics (us.st.com/stonline/index.shtml),
- Hyundai (www.hea.com),

or any other flash memory manufacturer.

SECRETS REVEALED

Now that we have obtained your car's original program and are ready to modify it, you may start to wonder about how complicated it looks. All you can see are thousands of numbers! If you care to know how they were created, read this chapter.

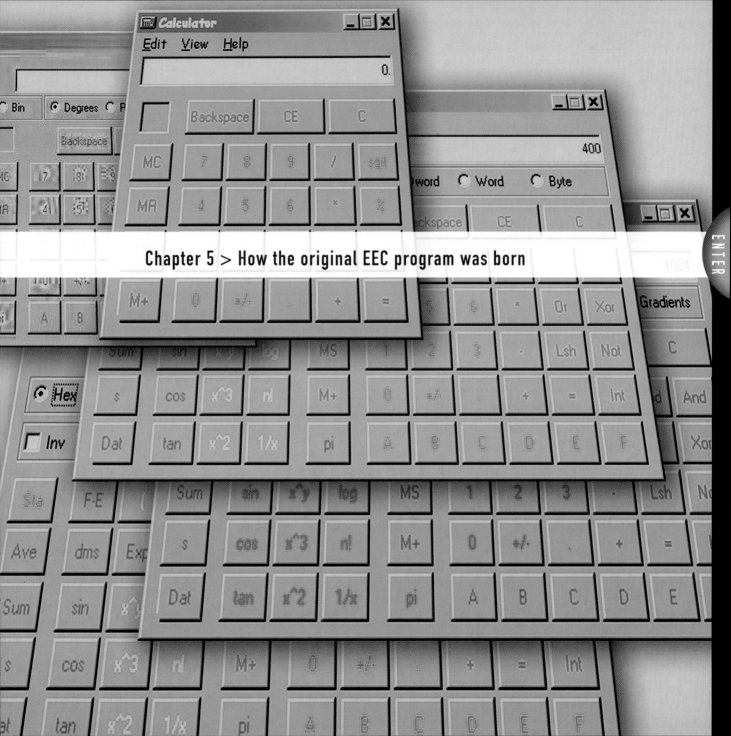

Chapter 5 > How the original EEC program was born

Chapter 5 > How the original EEC program was born

This chapter could also be named "The EEC Program - from Concept to Reality". It describes the steps the Ford design team had to take to create the data programmed into the Ford EEC memory. Understanding this concept helps us later to find the locations of data we have to modify in order to custom tune the performance of your Ford vehicle.

As already mentioned in the section about the microprocessor, the EEC program is nothing more than a detailed procedure describing to the microprocessor how to control the operation of a Ford engine.

The EEC program can be presented in various forms called computer languages. One way to visualize this, is that "How are you doing" in English means exactly the same as "Wie geht es Ihnen" in German – the message is simply translated to another language. The computer language translation process is called *compiling*.

Flowchart

The creators of the original Ford EEC program had to start with an engineering meeting. I imagine they each brought a yellow notepad and a large personalized coffee mug. They came up with a simplified diagram of the EEC program, describing the flow of the Ford engine control process. Engineers refer to it as a *flowchart*.

In layman's terms, a flowchart is a to-do list for the microprocessor. It describes the general flow of any procedure.

The following example shows a flowchart of a function: calculating the required fuel injection volume. This function might be called two hundred times per second, and the EEC microprocessor still has to handle multiple tasks (timing calculation, emissions monitoring, etc.)!

SECRETS REVEALED

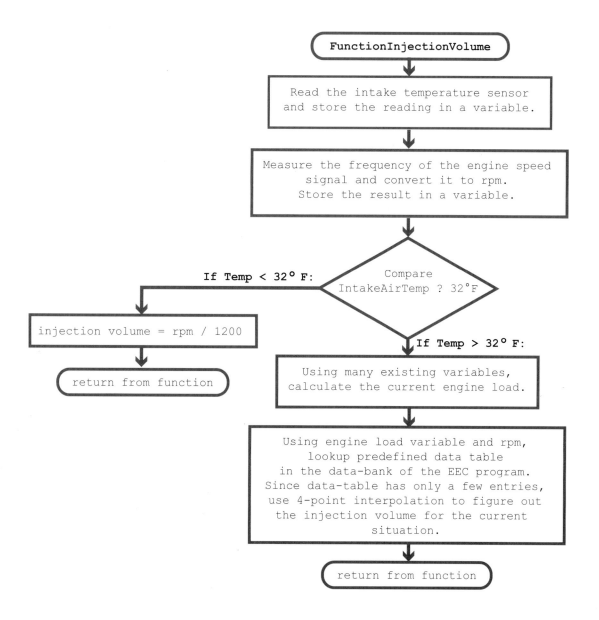

C programming language

Once the EEC flowchart was born, programmers took it to their cubicles and using it as a guideline, typed the EEC program on their PCs.

They hopefully wrote it in some efficient programming language, belonging to the category of *high level programming languages*, to save a lot of valuable time. Time in this case is definitely money, as programmers writing computer programs are among the highest paid!

Since we were not there and the source code of Ford engine control is confidential, we can assume that their choice was C programming language. It is one of the most popular high level programming languages for the job.

The previous example of a flowchart might be represented in a following C program:

```
FunctionInjectionVolume()
{
    IntAirTemp=IntakeAirTemperature();
    rpm=ReadSpeedSensor();

    if(IntAirTemp > 32F)
    {
        EngineLoad=CalculateLoad();
        InjVolume=LookupData(EngLoad,rpm);
    }
    else //too cold, use predefined value
        InjVolume=rpm / 1200;
    return(InjVolume);
}
```

It makes no difference whether C or some other high level programming language was actually used. The EEC microprocessor doesn't understand any high level programming language anyway. High level programming language is only used by programmers.

Notice how the above C language example is quite readable and in simple English. All programmers in the world write programs in English. Since it requires a vocabulary of only about thirty English words, don't assume a foreign programmer must understand your question "What time is it, please?". He might try to be polite and take his chance by answering "GOTO PRINT".

In layman's terms, programmers receive from engineers only a generic flowchart. Their job is to type a detailed program in special English, called high level programming language.

Assembly programming language

A program in C language might be easy to work with, but it cannot be programmed into the Ford EEC – the microprocessor doesn't understand it. The microprocessor, the brain of the EEC, understands only one language – *machine code*.

Imagine an ordinary calculator. Each time you press a key, you issue an instruction to the calculator's microprocessor to execute a small task. The calculator can execute only a very limited amount of tasks.

Similarly, the EEC microprocessor can execute only a limited amount of about 90 different tasks (instructions). It can add two numbers, subtract two numbers, compare two numbers, etc.

Each EEC instruction is assigned an identification number, called opcode. Opcodes are hard to remember, so engineers nicknamed them using short semi-English words.

For example, the instruction (or command) to add two bytes is `addb` and its opcode is `74`.

What is the reason to have both opcode and instruction name?

If you shop for groceries at the supermarket, each product is marked with a barcode. At the checkout counter, the barcode reader will read the barcode number and the printer will print the grocery name. The barcode number is for the computer and the grocery name printout is for the shopper.

If you really want to add two numbers together, you must specify both numbers to be added as well. To add five and three, the assembly code will be

```
addb   05   03 (use for humans)
```

and the matching machine code will be

```
74     05   03 (use for microprocessor)
```

The machine code is what gets programmed into the EEC ROM memory.

In layman's terms, high level programming language must be converted into machine code before it is programmed into the Ford EEC. When we read the EEC using a reader, all we get is the machine code. That is why, after the read, all we see is a lot of numbers on the PC screen.

Hexadecimal numbering system

Machine code, the only language of the EEC microprocessor, is the end result of the factory programmer's work. The machine code consists of hexadecimal, not decimal numbers.

Just look at a telephone keypad. There are ten different number keys, 0 - 9. The human decimal numbering system is based on 10; probably because we have ten fingers. Ancient mathematicians named them what they are today. They could be represented any other way, even as pictures of animals, if you wish. It is not logical; it is only an afreed to standard.

The EEC microprocessor doesn't understand numbers in the decimal format. Digital systems work on a system based on 16 numbers, called the hexadecimal system. In order to make their lives easier, mathematicians took the existing numbering system, 0 – 9, and just added the first six letters of the alphabet, A – F. (**Note**: Digital systems can really work on any numbering system based on two: 2, 4, 8, 16, 32, etc. For the purpose of this book, we are only using the above hexadecimal system).

In layman's terms, the machine code is always presented in a hexadecimal numbering system. The hexadecimal system can be converted into the decimal system just like one language can be translated into another. (Since the factory EEC program is always read out in the hexadecimal system, we need to convert it to decimal before tuning).

Example: Injection volume read out from the EEC is D2 (in hexadecimal). We want to simply increase it 10%. First we convert it to decimal (D2 = 210). We add 10% (210+ 10% = 231). The new injection volume must be converted back to hexadecimal (231 = E7). So, by changing the factory value of D2 to E7, we increased the injection value by 10%.

An easy way to convert one numbering system into another is to use a scientific calculator. Luckily, Microsoft Windows comes with one, which can save you the $20 it would cost to buy a calculator.

Start Windows Calculator

Run Windows Calculator by clicking on Start button, then select Programs > Accessories > Calculator.

Click on View on Calculator menu bar, and then select Scientific.

Convert decimal number to hexadecimal with Windows Calculator

Type in the decimal number to convert (for example 1024):

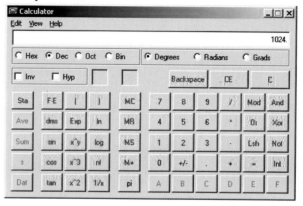

To see the hexadecimal equivalent of the number just entered, simply select Hex selector and the result is displayed:

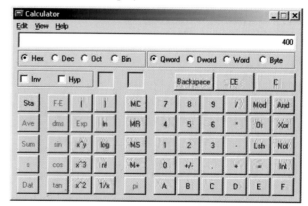

The above example converted decimal number 1024 into hexadecimal number 400.
To recognize whether a number is decimal or not, hexadecimal numbers are written differently. There are several standards to do that; one of them is to write character "h" after the number. This way the hexadecimal number 400 is written properly as 400h.

Convert hexadecimal number to decimal with a Windows Calculator

Click on Hex selector and type in the hexadecimal number to convert (for example FFFFh). The character "h" after the FFFF indicates that the number is hexadecimal and it is not to be typed into the Windows calculator.

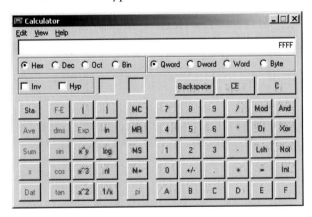

To see the decimal equivalent of the number just entered, simply select Dec selector and the result is displayed.

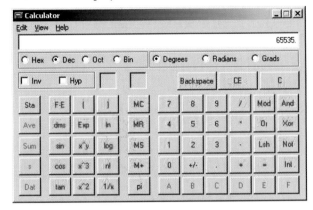

Executable program versus data tables

At this point we are familiar with all the steps of creating a Ford EEC program:

- Engineers made a flowchart.

- Programmers followed the flowchart and converted it into a programming language called C.

- C program was compiled into a machine code.

- The machine code gets programmed into the EEC memory.

When we read out the program from EEC, all we get is a binary machine code. We don't have

SECRETS REVEALED

the C program, or the flow chart. It stayed in the engineer's hands.

If engineers at the factory want to modify the performance and need to alter the fuel table, all they have to do is change the C program and recompile it into a machine code. That's it.

The aftermarket industry doesn't have this luxury. We have to find out, where in the ocean of numbers the fuel table is, and then carefully alter the numbers in it. Since we are modifying the machine code itself, we can program it into module memory without any compiling.

In layman's terms, data tables are spreadsheets in the EEC. Only the numbers in the data tables may be altered. Touching any numbers outside the tables is taboo! That is an executable program, the brain and reflexes of the EEC, and changing it will cause EEC not to work at all!

Data table location

At this point we understand that we may only alter numbers in the data tables. But, without the C program, how do we find them?

The most difficult and the most accurate way is to *disassemble* the entire EEC machine code. Disassembly, also called reverse-engineering, is the opposite process of what was done at the factory. In the factory they converted higher-level language into the machine code, disassembly converts the machine code back to higher language, which is much easier to understand. Disassembly is a painstaking process and takes an expert and many months to complete, especially when doing it for the first time.

We don't have that much time and must skip the disassembly process and find the data tables some other way.

Fortunately, the factory kept all the EEC programs relatively uniform and the data tables are always located in pretty much the same area.

For example, in the EEC-V, the data tables are always in the second bank*. One can quickly identify it – the first two bytes of the second bank are always 27 FE.

To learn more about data tables read the chapters that follow.

In layman's terms, the data tables are always in the second bank (second quarter) of the EEC program.

*(**Note:** If you are planning to design your own EEC reader, in order to read the second bank, the BS0 line must be high and BS1 line must be low. For more information about the bank reading, study the chapter about The EEC programa reader.)

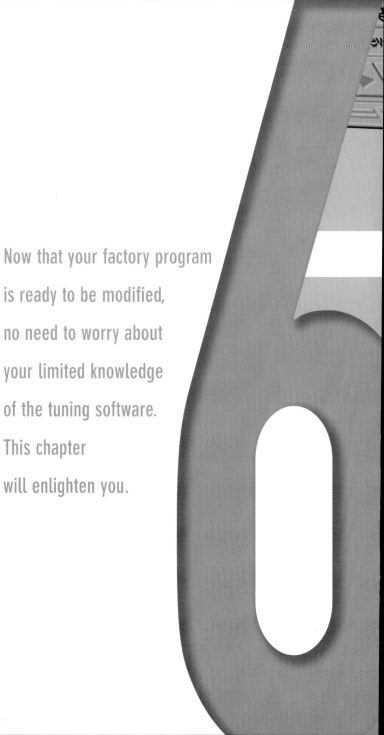

Now that your factory program
is ready to be modified,
no need to worry about
your limited knowledge
of the tuning software.
This chapter
will enlighten you.

Chapter 6 > Tuning software

contributed by Patrick Stadjel

FORD TUNING

Introduction

Why do we need to change what we are already given by the manufacturer?

It all started for me at the young age of 12 when I was racing motocross against competitors with the same store-bought racing machines. The only way to gain the power advantage was to select the proper carburetor settings and ignition-timing curve. We would make the adjustments based on fuel type, weather conditions and racing terrain. Whoever was able to do this well would have the horsepower advantage for that day.

Later on in life I graduated to my first real hot rod - a 1967 Ford GTA Mustang Fastback with a 351 Cleveland transplanted into it. The first thing I did was run out and buy a Crane Fireball camshaft - I "knew" that this would give me more power. How disappointed I was to learn how different the vehicle ran! It was not tuned for that camshaft; a lot more work was required. It needed a complete change of ignition curve and I even had to recalibrate the carburetor just to get the motor to idle properly. This required a lot of adjustment using only basic hand tools.

Then one day, while working "fat, dumb and happy" at my speed shop in New Jersey, a gentleman named Bob drove up in a 1986 Buick Grand National and changed my life forever.

Bob asked me if I could make his vehicle faster. I thought "no problem" and lifted the hood. To my surprise there was no distributor to adjust and the carburetor was missing. Since I was a hot rodder, I felt that all the same things would apply to this motor as well, but how do you change ignition timing and fuel delivery on a fuel-injected motor without a carburetor or distributor?

The answer was a new component called a computer; installed in all new vehicles. In the computer, there is a little silicone chip called an EPROM, which contains instructions about when to fire the spark plugs (ignition timing) and the length of time to activate the injectors (air-fuel ratio). A specialized chip reader can retrieve instructions from the chip and transfer them to a PC.

How do you take this information and change the timing and fuel tables? It took some trial and error to figure out how to arrange the information into maps for timing and maps for fuel. When we did, however, we were finally able to tune all of the newer vehicles. All the work paid off – Bob's Buick set an IHRA national record for the stock class!

Now that the introduction is out of the way, let's do some real work!

Tuning in Hexadecimal

Machine code is the native language of the EEC microprocessor. It is expressed in thousands of hexadecimal numbers. Most tuning software allows the tuner to view the EEC machine code and to modify any value at any location. There is no room for mistakes! One must know what is being changed and why – or damage to the engine could result.

If we had the original factory source code of the EEC program (probably written in a programming language called C), we would not need the tuning software – we could modify the data in the clearly specified data tables and compile the newly created source code back into machine code. This machine code would be programmed into a performance module and the tuning would be done.

Hex editing software, such as this one designed and used by DiabloSport, is the most common way to modify the code read from the EEC.

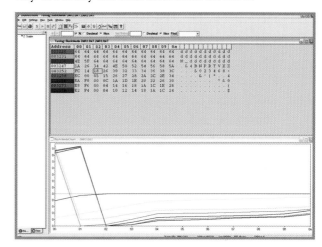

By using a scan tool (such as the Hellion diagnostic tool), this was determined to be a timing map. By changing the values on the PC and watching the effect of that change with the scan tool, we can see a change of timing. We still have to determine how to convert the hex value shown to a real timing value. Since we know both the actual timing degree from the scan tool display and the hexadecimal number from the PC display, we can figure out the formula to convert the hexadecimal number into a timing value.

Chapter 6 > Tuning software

FORD TUNING

In layman's terms, the PC replaces the screwdriver of yesterday, and the EEC code on the PC is the carburetor of yesterday. The tuning of the engine remains the same, only the tuning tools are new.

This was determined to be a fuel map. Again, by changing the values in the code and watching the effect of that change with the tool, we can see a change in the air/fuel ratio. This map must also be converted to real numbers by developing a formula.

of finding the tuning tables and converts the hexadecimal values to real numbers.

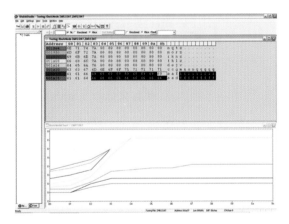

As you can see this is very cumbersome way to tune and very few people have the tools and experience to do this work. The newest method of user-friendly tuning, created by DiabloSport, had been a dream of mine for a long time. Their Revolution tuning software is a system that takes all of the hard work out

SECRETS REVEALED

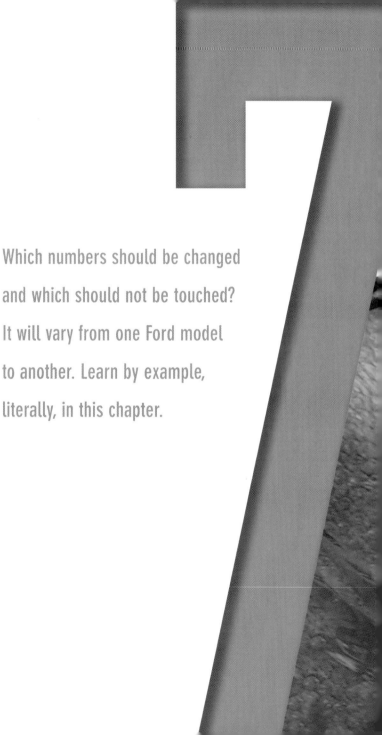

Which numbers should be changed and which should not be touched? It will vary from one Ford model to another. Learn by example, literally, in this chapter.

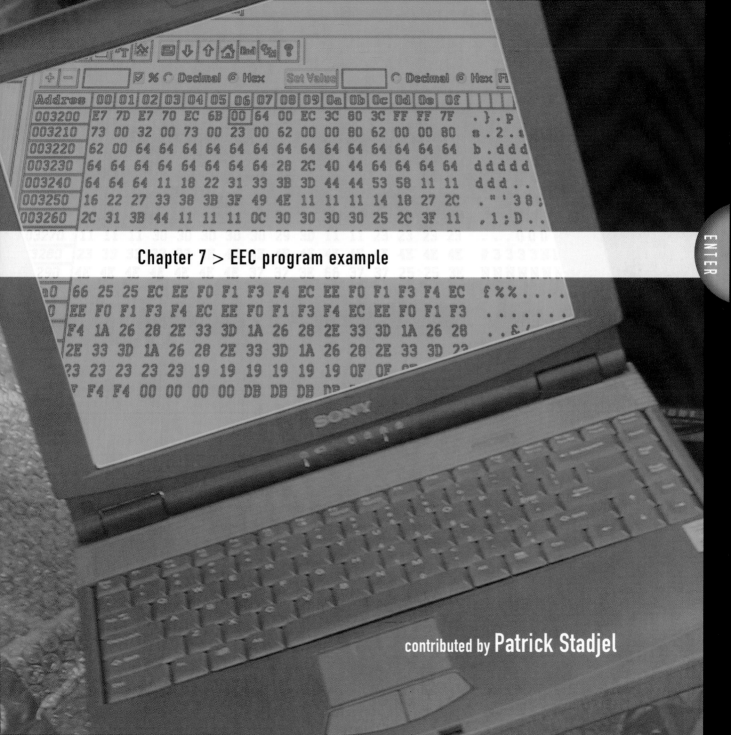

Chapter 7 > EEC program example

contributed by Patrick Stadjel

Chapter 7 > EEC program example

FORD TUNING

This chapter will take you through a tuning scenario using the DiabloSport Revolution tuning software. This is a very user-friendly way to tune a car. It is Windows based, menu driven and enables the user to choose exactly what to change and by how much.

The Vehicle: 1999 Mustang 4.6L Automatic

Equipment: Supercharged; 30-pound injectors; 3.73 gears off-road exhaust.

We are going to recalibrate the fuel system, the injection system and the engine management for the bigger 30-pound injectors. The gear ratio will also have to be calculated in order for the speedometer to display the correct speed. Since the supercharger adds more torque, the shift pressure should be raised to help keep the transmission from slipping. There is also an exhaust system installed which will require a change in the cat monitoring sensitivity.

First, we set the appropriate injector size. We type the proper size of the injector into the "select injector size" box. This will cause the car to idle properly.

Second, we set the appropriate gear ratio. We type 373 into the "gear selection" box. This will cause the speedometer to display the correct speed.

Next, we tune the engine on the dynamometer. We start with the timing tables. Patrick sets them to a low value, usually 10-8 degrees less than stock.

After this, one run on the dyno is done to monitor the air/fuel ratio. Looking over the test results, Patrick adjusts the fuel tables to get the ratios down or up to 11.8:1. For a supercharged street driven mustang like this one, a fuel ratio of 11.8:1 is ideal. It is possible to create even more power by running leaner, but the reliability is sacrificed.

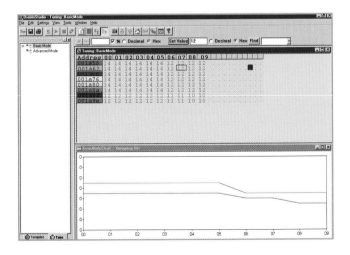

Once the fuel is set, we are back to the timing map. Patrick will slowly bring up the timing values until the desired horsepower and torque figures are reached. While this is done, we continuously monitor the air/fuel ratios to keep them within safe limits.

After the dyno session, the car is taken for road testing. This is a must since the conditions and loads vary from dyno to the real world. This can be simulated with a load bearing dyno, but it is better to road test the car. There will always be more changes made on the road.

The transmission is tuned at this point. The shift firmness is set so there is no slippage. The shift point is set based on the rpm peak horsepower read from the dyno printout. Patrick prefers to put the shift points 400-600rpm higher than the peak horsepower. This is done for the engine to stay in the power curve after shifting.

Tuning the timing table

Once the timing map is found, using the DiabloSport or similar tuning software with the emulator connected and the engine idling, you will notice a flashing square surrounding a number in the upper left corner. This is called a *trace*.

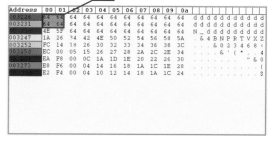

The values that are highlighted by the emulator are being accessed at that precise moment. If you rev the engine, you will notice that the trace will move across the map to the right further from its starting point.

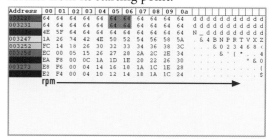

When the vehicle is put into gear, engine load will slightly increase. You will notice the trace drop down at idle, representing the new values being accessed and taking the load into consideration.

At this point, you can accelerate to the load point you want to change, and see exactly what values need to be modified for better performance.

For example, say there is a dead spot in the power curve when accelerating from 3000rpm to 5000rpm under full load. With the emulating tuning software you can feel the dead spot and see it in the highlighted area at the same time. This makes it easy to go to the specific spot and correct the problem by adjusting the numbers.

How can you be sure what values needs to be changed, and what to change them to? If you feel the dead spot, and see where on the map it is highlighted, you can then look at the values.

For example, while accelerating, you may notice the timing values incrementing like this:

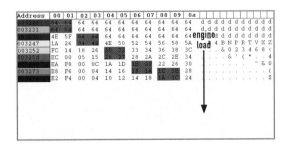

Notice that the flat spot has occurred only where the timing values are ramping up. It is possible that the lower timing values at this load and rpm are causing the flat spot. This curve may be necessary using a low-grade fuel to prevent detonation at this high load, low rpm area. If a better grade fuel is used, it may be possible to raise these timing values and remove the dead spot.

This is a commonly used method for tuning a timing table.

Tuning the fuel tables is very similar, but varies from application to application.

Tuning the transmission

Transmission tuning is slightly more complex. Patrick will tune the transmission under real conditions on the road. He believes tuning a transmission on a dyno is too difficult and it does not cover the vast amount of variables encountered on the road.

Setting the shift points on the transmission an average of 500rpm above the peak horsepower on the dyno curve is the first adjustment he will make. This is done so that after the shift, the rpm drops right back down to its sweet spot on the horsepower curve. Patrick does it for each gear, shifting 1-2, 2-3, and 3-4.

When that job is finished, he then adjusts the line pressure in the transmission. This affects how fast or how hard the transmission goes from gear to gear. On electronically controlled transmissions it is possible to set the pressure at different throttle settings. For example, at a cruise or light acceleration, you can keep the pressure very tame (you would not want to spill your hot coffee in your lap when pulling out of the driveway).

Once light throttle position pressure is set, Patrick will then really firm up the heavy load or W.O.T. (wide open throttle) shift points to minimize any transmission clutch slippage. He also reduces the time it takes the transmission to start and end the shift, thus minimizing the clutch slippage and reducing the transmission fluid temperature.

Once the transmission behaves the way he likes it, Partick will then remove the spark retard between shifts. This is how the stock program tries to make the transmission last forever. The way it works is that every time the computer sees a need for a shift, it will pull out a specified amount of timing right before the shift; thus reducing the horsepower and allowing for the shift to take place under less stress.

After the shift has occurred, the timing is then slowly added back until the next shift, but the transmission has already lost valuable power at a crucial moment. By eliminating this timing retard, you can gain quite a bit of horsepower between shifts. This power gain is difficult to see on a dyno run, since it is usually done in only one gear. However, it is very noticeable in the quarter mile and around town.

The next adjustment made to the transmission is usually done only for quarter mile drag racing. What Patrick does is this: once the car leaves the starting line and sets into the power curve (which can happen very early in the quarter mile run), he locks the torque converter to reduce the power lost through torque converter slippage to the rear wheels. He keeps it locked through all the gear shifts all the way through the quarter mile run.

Keep in mind for what type of use is the vehicle being tuned. In this case, Patrick tuned for a quarter mile race. However, when tuning for a 24-hour endurance racecar, high power and reliability are necessary. The power output has to be tuned to be more conservative.

Chapter 7 > EEC program example
FORD TUNING

A street car, on the other hand, needs to pass emissions and perform well under stop and go conditions. This is the most difficult type of tuning to do.

Obviously, there is a bit more to tuning than has been explained. These are only the basics. However, if you are interested in learning more, a school is being set up at DiabloSport. Besides, the team is working on another Kotzig Publishing book for advanced tuners, which will delve deeper into the secrets of tuning Ford automobiles.

SECRETS REVEALED

Even though most professional tuning is done on a PC in the comfort of an office, there are times when one must take the laptop computer into the car and tune live.

A wonderful companion to the tuning software, an interface box called ROM emulator is used and described in this chapter.

Chapter 8 > ROM emulator

This chapter describes a wonderful companion to the tuning software - an interface box called the *ROM emulator*.

Even though most professional tuning is done on a PC in the comfort of an office, there are times when one must take the laptop computer into the car and tune live. The laptop is really just a portable PC. Since the Ford EEC doesn't have any standard connection to a laptop, an interface box is necessary.

One function of the ROM emulator is to allow us to connect the PC to the Ford EEC, but there is something more important. As soon as the ROM emulator is plugged into the EEC, it disables the factory ROM chip located in the EEC (which contains the factory program) and replaces it with a duplicate chip (which can be tuned live on a PC).

In layman's terms, a ROM emulator is a modified performance chip which can be connected to a PC and tuned with a running engine.

Electronic tuning

We know already that electronic tuning is a process during which we alter a few bytes in the original Ford program using an ordinary PC.

Before we tune, we must read the original EEC program.

Obtaining the original Ford program:

• Turn off the ignition key and put it in your pocket, or else EEC damage might occur! (It was a winter day here in Florida, cold enough to wear a jacket. I followed these instructions religiously and put Patrick's ignition key in my jacket pocket. He was looking for it for a year, before I realized I still had it.)

• Unplug the Ford EEC from the vehicle.

• Read out the original program using an EEC reader.

• Save the readout on a PC.

• Plug the EEC back into the vehicle.

SECRETS REVEALED

We have a choice between two basic tuning procedures, non-live tuning (ROM emulator is not needed) or live tuning (ROM emulator is needed).

Non-live tuning process:
(See chapter about EEC program example)

• Alter a few bytes of the factory program (tune the program) on the PC using any tuning software (like DiabloSport Revolution).

• Transfer the altered data from the PC to a performance chip.

• Plug the performance chip into the EEC (make sure the ignition key is in your pocket, or else EEC damage might occur). This chip will disable the factory EEC ROM chip program and replace it with it's own.

• Start the engine.

• Verify that the new performance program is satisfactory (read the chapter on measuring performance). If not, continue tuning by repeating all of the above steps.

Live tuning process:
(See chapter about Steeda's Mustang tuned)

• Connect the laptop to the ROM emulator.

• Connect the ROM emulator to the EEC (turn off the ignition key and put it in your pocket, or else EEC damage might occur!).

• Start the engine.

• Alter a few bytes of the original program (tune the program) on the PC using tuning software. Any change of any value on the PC is immediately transferred to the ROM emulator and immediately affects the running engine – you are now tuning live!

ROM emulator block diagram

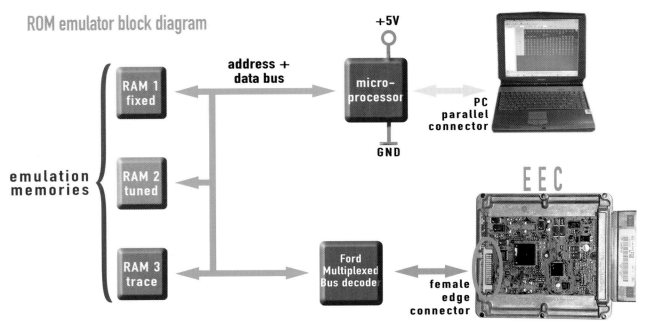

ROM emulator types

Poor man's ROM emulator

Scene of the "crime": Ford vehicle with engine running, ROM emulator plugged into the J3 connector of the EEC, a laptop plugged into the ROM emulator. the EEC is executing the program from the ROM emulator only, which is under total control of the laptop. There is clearly a live tuning going on here.

If you really think about it, the ROM emulator acts exactly as any performance module – it replaces the original Ford program. There is only one difference: the ROM emulator data can be altered live (real-time) by a PC.

How is it possible, that the externally plugged-in ROM emulator takes precedence over the factory ROM still soldered in the EEC?

The J3 connector on the EEC has one dedicated pin, used by the ROM emulator, which will disable the factory ROM (more details can be found in the chapter about performance chips).

In layman's terms, the basic ROM emulator is a modified performance chip, of which memory content can be tuned live with a PC.

SECRETS REVEALED

Poor man's (got a bonus and upgraded) ROM emulator

If you want to tune, and also have a "parachute" in case of emergency, you want one more function added to the above basic ROM emulator – a duplicate memory with emergency switchover. Why?

When you are at the beginning of the tuning session, downloading the original program from a PC into the ROM emulator, it takes at least a minute or two.

Then you can start the live tuning. If you change some byte to an incorrect value (for instance a very high timing value) the EEC will just blindly execute your request and possible engine damage could occur!

Before any damage from a very wrong byte change, the engine will probably ping and give you plenty of warnings that something is wrong, but there is simply not enough time to revert to the stock program. To make things worse, there is a human panic effect (delaying your reaction), or you might not know which byte was changed.

To solve this problem, a hardware upgrade to the emulator is needed. By doubling the emulator memory and keeping a duplicate factory program in the emulator, when an emergency happens, all you have to do is press a hot key on your laptop and the emulator will immediately revert to the factory program.

In layman's terms, the better ROM emulator has room for two programs – one stock and one performance. This allows for immediate switching back to the stock program in the case of an emergency or for comparison purposes.

(The real deal) ROM emulator

It almost looks like there is nothing else to wish for. But remember, a good salesman will always find something to sell you.

How about another upgrade to your emulator which will offer you a trace capability?

A trace is a special reverse-engineering capability of the ROM emulator, which allows monitoring of the activity of the EEC microprocessor. With appropriate training, it can helps us locate the data tables to be tuned.

How does it work?

The ROM emulator "knows", which locations in the program memory are being accessed by the EEC microprocessor. Of course it knows – it replaced the factory chip program. So every time the EEC microprocessor requests a data byte from any location, the ROM emulator will record that activity.

How to use a trace capability?

Let's say, for example, that you want to advance the engine timing in the area above 4,000rpm. You do not know exactly, which numbers to change. You believe, that you have the proper timing table on your laptop's display. You are either in the dynamometer room (read about it in the chapter Measuring performance) or in the vehicle being driven on the road (make sure that you are the passenger and someone else is driving). With the ROM emulator tracing function enabled, you will be able to see the highlighted numbers changing on the laptop's display in real time.

Monitor the current rpm with a handheld diagnostic scanner. You should be able to identify the timing values in the requested rpm range.

 In layman's terms, tracing means that the ROM emulator is spying on the EEC microprocessor and helps us locate the tuning tables.

DiabloSport tuning school

Be advised, it takes some experience and talent to be able to reverse-engineer the tuning tables. There is definitely logic in the process and common sense is a necessity; what helps tremendously is training. If you are interested in getting properly trained in this field, contact your DiabloSport dealer. You will find more information on the website at www.diablosport.com.

DiabloSport ROMulator

There are not many automotive ROM emulators on the market. The ones we've seen are all made in Europe. DiabloSport decided not to import them and designed and manufactured its own ROMulator here in the U.S.A.

SECRETS REVEALED

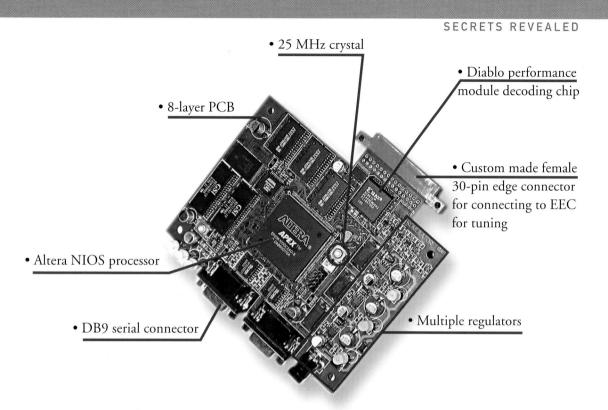

- 25 MHz crystal
- Diablo performance module decoding chip
- 8-layer PCB
- Custom made female 30-pin edge connector for connecting to EEC for tuning
- Altera NIOS processor
- DB9 serial connector
- Multiple regulators

Normally, there are many chips in a good ROM emulator. DiabloSport engineering selected a new, very modern component – Altera's NIOS, a soft-core microprocessor.

Soft-core means that a microprocessor design is purchased as software (not as a chip) and can be customized in-house using an ordinary PC. DiabloSport customized the NIOS by including a lot of dedicated functions, which tremendously simplified the hardware design.

The ROMulator is running its own operating system, called KROS. (KROS was made so well, that even the creator of the NIOS processor, Altera Corporation, www.altera.com, asked for it. That jump-started a new DiabloSport sister company – Shugyo Design Technologies, www.shugyodesign.com).

You are done with PC tuning and want to go back to your beloved Ford and drive it! This chapter describes the way your work gets transferred from the PC to the car.

Chapter 9 > Performance chip

Is it a chip or module?

Call it what you will, it doesn't matter, both names are commonly used. Street slang often calls it a *performance chip*, referring to a practice on non-Ford vehicles, where the tuned program is often delivered on an ordinary EPROM chip. The factory EPROM is removed and replaced with an aftermarket chip.

This is not possible on any Ford vehicle, where the ROM is non-standard and not available to the public. Instead, it must be replaced with a performance module.

The fortunate thing is that not only does the original ROM not have to be de-soldered, but also the EEC doesn't have to be opened at all! There is a convenient opening on the rear end of every EEC, exposing a 30-pin edge connector (J3 connector), into which the module can be plugged.

Layman, you might want to skip the remainder of this section.

How can the externally plugged-in chip disable the factory ROM, still soldered in the EEC?
All the vital pins of the internal ROM are routed to the J3 edge connector, where the performance module connects. That makes the performance module practically piggy-backed onto the original ROM. Now that alone wouldn't work, because both the original and the tuned ROM would be active at

SECRETS REVEALED

the same time; having different data, it would cause the EEC not to function. Engineers at the factory thought about that and provided an Internal_ROM_disable control pin on the same edge connector. When nothing is plugged-in externally, the Internal_ROM_disable is not activated and the EEC microprocessor is free to access the internal ROM. The performance module has an electrical connection to this pin and activates the Internal_ROM_disable signal. That disables the internal ROM while the performance module is plugged in. Because the performance module is in parallel to the internal ROM address- and data bus, the EEC microprocessor has actually no idea that the data comes from an external location.

How it works

A typical aftermarket performance module contains two ICs, or chips (as normal folks call them). Both off-the-shelf chips together equal what Ford and Intel engineers enclosed in their single chip proprietary ROM.

The first IC on the performance module is decoding the Ford Multiplexed Bus into a standard non-multiplexed address- and data bus. Once that is done, almost any ROM, whether EPROM or flash ROM, can be used to hold the tuned program; the second IC is just that.

In the performance modules designed years ago, there would only be an IC socket instead

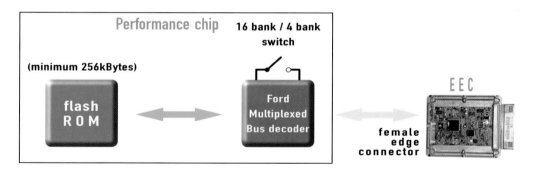

of ROM. To finalize the module assembly, the dealer would have to program a standard EPROM with the tuned program, and plug it in the socket. When a need to change the program occurred, the EPROM chip would have to be removed from the socket and replaced. The dealer could, optionally, erase the original EPROM with ultraviolet light (a lengthy procedure taking several minutes) and reuse the chip.

A modern version of an EPROM is a flash ROM. A flash ROM can be erased and reprogrammed an unlimited amount of times without removal from the module.

1-bank and 4-bank performance chips

Most performance chips on the market are not programmed with the entire EEC program, spanning up to 4-banks. They contain only the data portion, which typically is the size of only 1-bank. These modules are often called data-bank modules.

A data-bank performance module is designed in such a way that it is only activated when the EEC microprocessor needs to read data tables (as soon as the performance module decodes an attempt to read a data location, it "wakes up", disables the EEC internal ROM memory and replaces it with its own)*. Data-bank performance modules are usually easier to manufacture, but have one disadvantage. They only function until your Ford's vehicle EEC gets reprogrammed at the dealership (if there is an update for your vehicle available, dealerships will reprogram your EEC as a courtesy).

In another words, if the module was programmed with, for example, the data portion of a NVK4 program, it must be plugged into an EEC containing NVK4 program as well.

The owner should remove the performance module before any service, as it is always better to have the vehicle in stock condition so as not to confuse the technician. Unless warned, the technician will follow the normal procedure and reprogram the EEC with the latest update program available. This is how your EEC, originally having the NVK4 program might suddenly become an NVK7! (How can the technician change the EEC program without taking the EEC apart? Read the chapter about Handheld performance tools). Unfortunately, technicians don't always

---

*Editor note: ??…lost here…maybe cause it's 1:00AM and getting sleepy…time for some coffee, be right back…

replace the sticker on the EEC unit reflecting the new program name.

What happens next? The vehicle owner will pick-up the vehicle (perhaps only after an oil-change), plug his or her performance module back into the EEC – and the vehicle won't start! After unplugging the module, the vehicle will run just fine. Did the module get destroyed somehow? No, it only has a mismatched program in it. What he has to do is call his performance dealer and have the performance module reprogrammed (re-flashed).

1-bank (or data-bank) performance modules can give us a headache in a slightly different scenario as well. Even if we never had a performance module installed before, we might not know the correct EEC program code in our vehicle. All we can do is read the sticker on the EEC to our mail-order performance dealer. If they send us a 1-bank module and our vehicle was previously re-programmed with a different code, the car will not start. Most likely, we will blame the module manufacturer for a defective product; when, in reality, a new sticker was not put on the EEC box to indicate an update.

The 4-bank chip contains the entire EEC program and eliminates the above complications. Another benefit for performance dealers is the reduced inventory of EEC program codes. If the customer orders one of many versions of an EEC program for the same engine family, the dealer can mail the currently best performing one and rest assured that the vehicle will start!

A data-bank performance module is designed so that it is only activated when the EEC microprocessor needs to read data tables (as soon as the performance module decodes an attempt to read a data location, it "wakes up", disables the EEC internal ROM memory and replaces it with its own).

A 4-bank performance module is designed in so that it is activated all the time (it disables the EEC internal ROM memory).

Even if the EEC is re-programmed by any repair technician to any code, the original 4-bank performance module will work and the vehicle will start. Whew!

Programming a performance chip

At this point we understand that an aftermarket performance module should contain a modified factory EEC program. When plugged into the vehicle, it will replace the factory program and the vehicle will behave according to our modifications.

Programming of the performance module is done using dedicated equipment available from all performance module makers.

If you would like to know more about programmers, read the chapter EEC program reader. You will realize, that to read out a program from the EEC unit is a very similar process to programming a performance module – both require generation of Ford Multiplexed Bus signals.

Installing a performance chip

Before touching anything, remove the ignition key and put it in your pocket, assuring the ignition is in off-position. Failure to do so may result in a damaged EEC and/or performance module.

STEP 1: Remove the EEC from the vehicle.

STEP 2: Remove the J3 test port cover from the back of the EEC.

STEP 3 (most crucial!): Use a swab to wipe off the grease from the J3 edge connector. It also might be necessary to scrape off the solder mask (usually green color) from the J3 edge connector to expose its metallic contacts. One could use a Scotch Brite® Pad or very fine sand paper. In the worst case, even a small screwdriver will do. Don't forget to do it on both sides of the EEC computer board! Well-cleaned contacts are uniformly shiny.

STEP 4: Verify that the metallic contacts are free from dust and metallic particles. Clean if needed.

STEP 5: Insert the performance module all the way in. Make sure it is straight and makes a firm contact.

SECRETS REVEALED

STEP 6: Install the dust protective label.

STEP 7: Reinstall the EEC back into the vehicle.

STEP 8: Start the vehicle and enjoy!

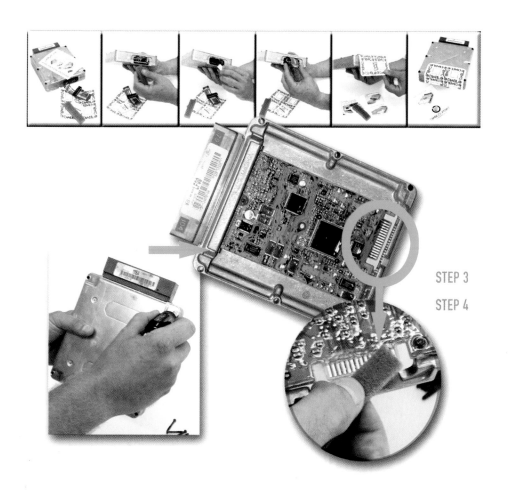

STEP 3
STEP 4

Performance chip and your vehicle's warranty

One very natural and commonly asked question is whether a performance module will void your vehicle's original warranty.

It absolutely shouldn't. You might be told the opposite, but there is a law protecting vehicle owners from just this kind of fraudulent behavior practiced by some automotive dealers. The Magnusson - Moss Warranty - Federal Trade commission Improvement Act of 1975 states, that aftermarket equipment that improves performance does not void a vehicle manufacturer's original warranty, unless the warranty clearly states that the addition of aftermarket equipment automatically voids the vehicles warranty or that the aftermarket device is the direct cause of the failure. The easiest way to check this is to look in your owner's manual under, "what is not covered".

Those so inclined can verify this at *http://www4.law.cornell.edu/uscode/15/ ch50.html/PC50* and at *http://www.diablosport.com/legal.html*.

By law, the installation of a performance module does not void the warranty. However, when you need to take your vehicle in for service or repairs, it might be a good idea to remove the module for the following reasons:

1) Anytime a diagnostic scanner is used, the vehicle's computer needs to be in stock condition. In order for the diagnostic scanner to accurately perform its task, it must be able to communicate with the stock computer program. With the module installed, the diagnostic scanner will not be able to do so. Failure to restore the stock tuning can result in unnecessary repairs.

2) In the case of a problem, it will confirm whether there is a problem with the module or with the vehicle.

3) Each individual dealer can, at their discretion, set up their own warranties and policies that would void parts of, or the entire, warranty by the addition of a non-Ford part.

DiabloChip

Unlike other performance modules, the DiabloChip plugs entirely inside the EEC, which allows you to "seal" off the control unit. This prevents the chances of dust or moisture from getting inside. For some vehicles, this also allows you to reinstall the EEC back into its original location without having to make any modifications.

The times are changing and the future is here — NOW! The tuning principles are still the same, but the new Fords are "missing" the familiar opening for the performance module in the computer box. Read this chapter to familiarize yourself with this new technology.

Chapter 10 > Handheld performance tool

Chapter 10 > Handheld performance tool

So far we have done all the necessary steps required to create a tuned Ford file. We:

- unplugged the Ford EEC from the vehicle
- read out the original program using an EEC reader
- saved the readout on a PC
- plugged the EEC back into the vehicle
- altered a few bytes of the factory program

We went even further and programmed the result of our tuning into the performance module. It worked well!

How does this process differ on new vehicles "missing" the familiar opening for the performance module in the EEC computer box?

Really, there is very little difference. The work is the same; only the tools required get more complicated. The tuning steps remain the same as the above list, all the way to the point of altering the original EEC file on the PC.

What is different is that the result of our work cannot be programmed into the module anymore – we need a handheld performance tool.

In layman's terms, since the new Ford vehicles will no longer accept the performance module, we need to reprogram them using a handheld performance tool.

OBDII communication

A handheld performance tool looks just like an ordinary diagnostic scanner, used by repair technicians to troubleshoot Ford vehicles.

It also works like a scanner, with the exception of different software loaded into it at the factory. That means, that any scanner manufacturer could turn their scanner into a handheld performance tool.

Every Ford manufactured after 1995 must comply to the OBDII standard and have a 16-pin female OBDII diagnostic plug located somewhere near the driver's seat. Ford generally uses pins 2 and 10 of the OBDII connector for the diagnostic. Those two pins are connected to the *differential bus*, Bus+ (OBDII pin 2) and Bus- (OBDII pin 10), and are routed to every control unit in the vehicle.

The OBDII standard allows all vehicle manufacturers to choose from various hardware interfaces and various communication protocols (which creates quite a mess out there…):

- J1850 PWM
- J1850 VPW
- ISO 9141
- CAN

If you want to learn basics about these, read the chapter OBDII diagnostic. If your interest is deeper than that, and you want to learn all about the OBDII hardware and software, get our next book OBDII Diagnostic Secrets Revealed. (See our website www.kotzigpublishing.com).

Ford J1850 PWM

Ford uses *J1850* PWM (Pulse Width Modulation) encoding on the diagnostic bus. That means, unlike a typical PC serial communication, where the voltage level of the signal dictates the logical value of the data (Logical 0 is defined as +10 Volts; Logical 1 is defined as -10 Volts), Ford control units ignore the voltage level. Instead, they monitor every pulse width, similar to a Morse code.

If you want to "see" an OBDII communication, connect an oscilloscope to the Bus+ wire. Set the oscilloscope to monitor +5 Volt signal levels and look at the signal waveforms.

FORD TUNING

Logical 0 is defined as a ground signal changing to +5 Volts, staying there for 7 Time Units, then switching back to ground and staying there for 17 Time Units.

Logical 1 is defined as a ground signal changing to +5 Volts, staying there for 15 Time Units, then switching back to ground and staying there for 9 Time Units.

In layman's terms, in Ford vehicles, the only difference between logical 0 and logical 1 is the duration of the pulse – hence the name Pulse Width Modulation.

Principle of operation

Engineers at the Ford factory do not use a performance module - they use an OBDII handheld tool.

Once the OBDII scanner tool is connected, it will transmit a message to the vehicle. The EEC microprocessor will receive the message and verify it. If the message is valid, the EEC will respond.

A typical scanner message looks as follows:

```
C4 10 F1 23 09 FF 06 8D
 |  |  |  |  |  |  |  |
 |  |  |  |  |  |  |  `- Message checksum
 |  |  |  |  `-------- Address FF06
 |  |  |  `-------------- Request data command
 |  |  `----------------- Who sent the message (scanner)
 |  `-------------------- For whom is the message (EEC)
 `----------------------- Required header byte
```

In the message above, the scanner is requesting data from the EEC memory at address FF06. This is another way to read out the EEC ROM memory, without the use of J3 port!

The EEC should respond with a message (data bytes will vary):

```
C4 F1 10 63 FF 06 56 4E 41 41 92
 |  |  |  |  |  |  |  |  |  |  |
 |  |  |  |  |  |  |  |  |  |  `- Message checksum
 |  |  |  |  |  |  |  |  |  `---- 4th Data byte
 |  |  |  |  |  |  |  |  `------- 3rd Data byte
 |  |  |  |  |  |  |  `---------- 2nd Data byte
 |  |  |  |  |  |  `------------- 1st Data byte
 |  |  |  |  `------------------- Address FF06
 |  |  |  `---------------------- Response to the request data command
 |  |  `------------------------- Who sent the message (EEC)
 |  `---------------------------- For whom is the message (scanner)
 `------------------------------- Required
```

SECRETS REVEALED

The scanner can transmit many kinds of messages, described in the OBDII literature. If an invalid message is transmitted to the EEC, it will simply ignore it.

The EEC also understands non-OBDII messages. Where are they described? Not where we can reach it!

Engineers at the Ford factory created a handy back-door feature – the EEC reprogramming ability. Before it can be used, the EEC must be unlocked.

To unlock the EEC, we must calculate the unlocking Key. To calculate the Key, we must obtain a Seed (a special number) from the vehicle. The Seed will vary from vehicle to vehicle.

Request the Seed as follows:

```
C4 10 F1 27 01 AA
 |  |  |  |  |  |
 |  |  |  |  |  `-------- Message checksum
 |  |  |  |  `----------- Request unlocking Seed
 |  |  |  `-------------- Who sent the message (scanner)
 |  |  `----------------- For whom is the message (EEC)
 |  `-------------------- Required header byte
```

A response to the Seed request message will be (Seed will vary vehicle to vehicle):

```
C4 10 F1 27 02 12 34 5A
 |  |  |  |  |  |     |
 |  |  |  |  |  |     `---- Message checksum
 |  |  |  |  |  `---------- unlocking Key = 12 34
 |  |  |  |  `------------- Unlocking with a Key
 |  |  |  `---------------- Who sent the message (scanner)
 |  |  `------------------- For whom is the message (EEC)
 |  `---------------------- Required header
```

Now comes the tough part. The equation to calculate the Key, unfortunately, varies car-to-car, model-to-model and year-to-year! For illustration purposes only, here is what the scanner has to send to unlock the vehicle for the above Seed:

```
C4 F1 10 67 01 AB CD EF 6C
 |  |  |  |  |        |  |
 |  |  |  |  |        `- Message checksum
 |  |  |  |  `---------- Seed = AB CD EF
 |  |  |  `------------- Sending the Seed
 |  |  `---------------- Who sent the message (EEC)
 |  `------------------- For whom is the message (scanner)
 `---------------------- Required header byte
```

Once the EEC is successfully unlocked, the performance tool sends the data to be written using standard OBDII messages.

The data being sent is, naturally, the same machine code we used to program the performance module in previous chapters.

DiabloSport Predator

DiabloSport upgraded the existing technology of the OBDII *Hellion* diagnostic scanner (www.eobd.com), and created a handheld performance tool, called the Predator (as they joke about it, **P**erformance **R**eady **E**lectronic **D**iagnostic **A**lteration **T**ool **O**BDII **R**eady).

Predator is a combination of a powerful diagnostic scanner and a user-friendly performance programmer.

DiabloSport designed the Predator exactly the way we need it – we can ignore the intricacies of EEC reprogramming, described in this chapter.

In layman's terms, the Predator can be programmed as a performance module, and then carried over to the car. Using a user-friendly graphic interface, our tuned file can be written to the EEC. Isn't that great?

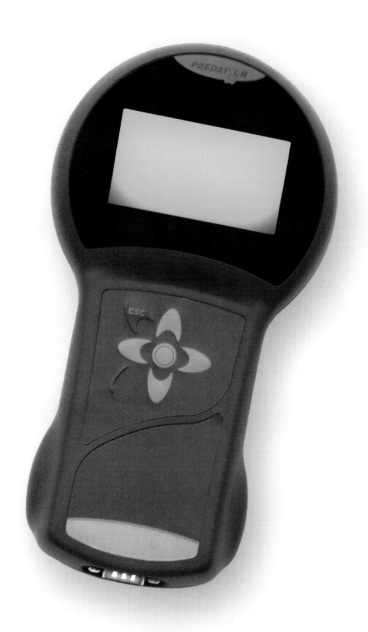

SECRETS REVEALED

Read this chapter and learn about various types of dynamometers.

11

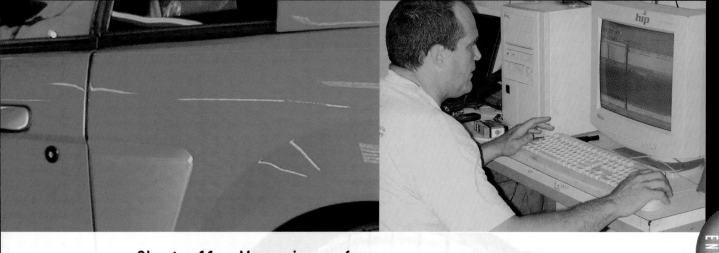

Chapter 11 > Measuring performance

Chapter 11 > Measuring performance

FORD TUNING

A *dynamometer* is a machine used to measure horsepower and torque.

Dynamometers can be set up in two categories:

- An Engine Dynamometer is used to monitor and measure the performance on an engine alone.

- A Chassis Dynamometer allows the entire vehicle to be tested as one unit, taking driveline power loss into consideration. There are four-wheel drive (4 rollers), two-wheel drive (2 rollers) and even motorcycle chassis dynamometers.

Engine dynamometer

Most common during engine development, allows for easy access to engine and testing under controlled conditions.

Engine dynamometers typically measure power at the flywheel, as this is the power measurement usually given by vehicle manufacturers. This number is always higher than a chassis dynamometer due to the fact that loss of power through the drive train is eliminated.

Common uses for an engine dynamometer:

- Research and development – fine tuning of high performance engines

- Long term testing with extremely accurate measurements

- Stand alone engines without vehicles

Chassis dynamometer

Most common among tuners, allows the entire vehicle to be tested in a stationary position. The advantages of a chassis dynamometer are that there is no need to remove the engine from the vehicle and that it measures the true power at the wheels, taking drive train power loss into consideration.

Typically constructed of large rollers driven by the vehicle wheels. Rollers have sensors that calculate horsepower and torque.

The most commonly used types of chassis dynamometers:

- Load bearing - Has the ability to apply and measure resistance upon the rollers. Useful to monitor power behavior, emissions, fuel consumption and air/fuel ratios under various loads.

- Inertia - The most common type of dynamometer and also the least expensive. This type does not have the ability to apply resistance upon the rollers. The main use of this type of dynamometer is to measure peak horsepower and torque at the wheels.

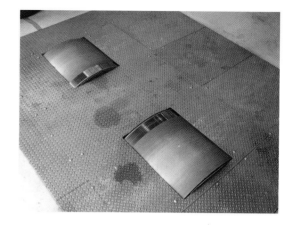

Common uses for a chassis dynamometer:

- Quick and easy monitoring of power

- Complement the engine dynamometer as a final test

- Monitor driveline power loss

- Acquire accurate information on individual accessories

- Monitor noise, fuel consumption, air/fuel ratios and emissions

This chapter is an excerpt from the OBDII Diagnostic Secrets Revealed book. It is an introduction to the world of contemporary diagnostic tools. Tuners can take advantage of taking various measurements from a running engine and monitor effects of their tuning in real time.

Chapter 12 > OBDII diagnostic

contributed by Peter David

Chapter 12 > OBDII diagnostic

FORD TUNING

On-Board Diagnostic II (OBDII) is a standard that almost everyone today has either seen or used. The funny thing about it is that most people don't even know that OBDII is influencing their vehicle. There are millions of vehicles on the road with OBDII, influencing all our lives with cleaner air. It might be today's least known, but most used, automotive concept.

OBDII compatibility was incorporated on a few models in 1994, including the Ford 3.8L V-6. In 1995, more models were added. Then in 1996, federal regulations mandated OBDII to be used on all new cars and light trucks (domestic and imported) sold in the United States.

How did OBDII come about? In 1989, because of the continued failure of emission systems dating back to 1966, the *Environmental Protection Agency* (EPA, www.epa.gov) finally got together with the California Air Resources Board (CARB, www.arb.ca.gov/homepage.htm) and perfected the practices that CARB had been doing for years. They wanted a better way to detect engine performance problems that caused emissions to rise. The goal was cleaner air and a better environment. This system would not replace the emissions testing, but would act as an On-Board Diagnostic emissions monitor.

The concept of OBDII has been around for years. In the past, each manufacturer used their own systems and controls. To resolve this, the Society of Automotive Engineers (SAE, www.sae.org) proposed several standards. The birth of the OBD was marked when the CARB mandated many of the SAE's standards for emissions in California on 1988 and later vehicles.

The original OBD system was not complex at all. It monitored the Oxygen sensor, Exhaust Gas Recirculation (EGR) system, fuel delivery system and the Engine Control Module (ECM, referred to as EEC in Ford) for excessive emissions. It did not require any uniformity from the manufacturers. Each vehicle manufacturer had their own procedure for monitoring emissions, as well as diagnosing the system when the emissions were out of the normal range. The emission monitoring systems were not working efficiently since the manufacturers were designing them for vehicles already in production. Vehicles not originally designed for the emission systems failed drastically and the manufacturers were not necessarily complying with regulations. They did (more or less) what CARB told them to, but nothing more. Imagine if you were an independent repair facility. You had to have a unique diagnostic tool, manuals for codes, and repair manuals for each manufacturer. Vehicles were not getting repaired properly, if at all.

The United States Government was besieged on all sides, from independent repair facilities to lobbyists for clean air. EPA was asked to step in. They took the standards of the SAE and the ideas of CARB and created an extensive list of procedures and standards. By 1996, all manufacturers had to comply in order to sell vehicles in the United States.

The second generation of On-Board Diagnostics was birthed, with the name of OBDII.

The OBDII standardization included the following:

- One centralized diagnostic connector with pins assigned specific functions

- Standardized diagnostic communication protocols

- Standardized set of Diagnostic Trouble Codes (DTC)

- Standardized component and emissions terminology

One centralized diagnostic connector

The diagnostic connector (OBDII calls it the Diagnostic Link Connector - DLC) must have a centralized location in the vehicle. It must be within 16 inches of the steering wheel. The manufacturer can place the DLC in one of eight possible places predetermined by the EPA. The DLC's main function is to allow a diagnostic scan tool to communicate with OBDII compliant control units. Each pin has its assigned definition. Assignment of many of the pins is still left up to the manufacturer, but those pins are not meant for OBDII compliant control units. They can be for Supplemental Restraint Systems (SRS), Anti-Lock Brake Systems (ABS) just to name a few.

The standard OBDII connector

Pin 9 – Proprietary

Pin 10 – J1850 Bus-

Pin 11 – Proprietary

Pin 12 – Proprietary

Pin 13 – Proprietary

Pin 14 – CAN Low (J-2284)

Pin 15 – ISO 9141-2 (L Line)

Pin 16 – Battery Power

Pin 1 – Proprietary

Pin 2 – J1850 Bus+

Pin 3 – Proprietary

Pin 4 – Chassis Ground

Pin 5 – Signal Ground

Pin 6 – CAN High (J-2284)

Pin 7 – ISO 9141-2 (K Line)

Pin 8 – Proprietary

J1850, CAN and ISO 9141-2 are protocol standards developed by *SAE* and *ISO*. The manufacturers have their choice of these standards to use for their diagnostic communication.

Each standard has a specific pin to communicate on. For example, Ford products communicate on pins 2, and 10. GM products communicate on pin 2. Most Asian and European products communicate on pin 7, and some also on pin 15.

Which protocol is used, makes no difference to understanding OBDII. The message exchanged between the diagnostic tool and the control unit is always exactly the same, only the way it is transmitted differs.

One standardized connector, with one shape, in one location, makes it easier and cheaper for the repair shops. Now they don't need 20 different connectors or tools for 20 different vehicles. In addition this saves time, since the repair shops won't have to hunt down the location of the connector to hook up the tool.

Standardized diagnostic communication protocols

As seen above, there are several different protocols that OBDII recognizes. At this point, we only need to discuss two protocols, J1850 and ISO9141, which directly affect all the vehicles in the United States.

All the control units in the vehicle are connected with a cable (called a diagnostic bus), creating a network. We can connect a Diagnostic Scan Tool to the diagnostic bus. The tool will send out a signal to the specific control unit with which it wants to communicate. The control unit will respond. Communication will continue until the tool terminates communication or the tool is disconnected.

For instance, the Diagnostic Scan Tool will ask the control unit "What are your faults?" The control unit will answer appropriately. With that simple communication exchange, we have just followed a protocol.

In Layman's terms, protocol is a set of rules that must be followed in order for a network to complete a communication.

Classification of a Protocol

SAE has defined three distinct protocol classifications: Class A, Class B, and Class C.

Class A is the slowest of the three and can be as high as 10,000 bytes per second or 10Kb/s. The ISO 9141 standard uses the Class A protocol.

Class B is ten times faster and supports communication of data as high as 100Kb/s. The SAE J1850 Standard is a Class B protocol.

Class C supports communication performance as high as 1Mb/s. The most widely used vehicle-networking standard for Class C is Controller Area Network (CAN). Higher performance communication classifications from 1Mb/s to 10Mb/s are expected in the future. Classifications like *Class D* can be expected as bandwidth and performance needs go forward. With Class C, and the futuristic Class D protocols, we will be able to use fiber optics as cabling for the network.

J1850 PWM protocol

J1850 comes in two different flavors. The first is a high speed 41.6 Kb/s Pulse Width Modulation (PWM). This is used by Ford, Jaguar and Mazda. Ford was the first to use this type of communication. The communication uses two wires, pins 2 and 10 of the diagnostic connector.

J1850 VPW protocol

The other J1850 alternative is the 10.4 Kb/s Variable Pulse Width (VPW). This protocol is used by both General Motors (GM) and Chrysler. It is very similar to Ford's protocol, but the communication is much slower. It uses one wire, pin 2 of the diagnostic connector.

ISO 9141 protocol

The third protocol is the ISO 9141, defined by the International Standard Organization (ISO). This standard is used by most European, Asian, and some Chrysler vehicles.

It is not as complex as the J1850 standards. While the J1850 protocols require use of specialized communication microprocessors, ISO 9141 uses standard off-the-shelf serial communication chips.

In layman's terms, OBDII uses a standardized diagnostic communication protocol, because the EPA wanted a standard way for repair shops to diagnose and repair vehicles properly. They wanted this without the expense of proprietary equipment.

A standardized set of Diagnostic Trouble Codes (DTC) for ALL manufacturers

When the engine management has detected an emission related problem, the Check Engine light will come on. OBDII calls the light a Malfunction Indicator Light (MIL).

The MIL will not necessarily come on when a fault first happens. The fault must happen on several occasions, called drive cycles. A drive cycle is when a vehicle is started cold and driven to normal operating temperature (with coolant temperature below 122 degrees F and the coolant and air temperature sensors within 11 degrees of one another). During this process all the on-board emission monitor tests must be completed.

Different vehicles have different engine sizes, and the drive cycle is slightly different for each. On an average vehicle, if a fault is seen in 3 drive cycles then the MIL will come on. On the other hand, if the fault is not seen in 3 drive cycles, then the MIL will go out.

The fault remains stored in the EEC, and can be retrieved with a scan tool. There are two states that the fault can be in: Stored or Active. Stored is when the fault has been detected, but the MIL is not on, or was on and went out. Active is when the fault is present, and the MIL is on.

OBDII calls a fault a Diagnostic Trouble Code (DTC). A DTC is made up of a combination of 1 letter and 4 digits. The table below shows what each character means. It helps to quickly identify the general area of the problem without knowing the exact description of the code.

As you can see, each character has a meaning. The second character is the most controversial – it shows who defined the code. Value 0 (commonly known as "P0" codes) means it is a generic, publicly known SAE fault code. (See below for a complete list of generic fault codes). Value 1 (commonly known as "P1" codes) means it is a proprietary fault code, defined by the vehicle manufacturer. Most scanners are unable to recognize the description or text of the P1 codes. However, Hellion by DiabloSport (*www.eobd.com*) is one of the few that can recognize most of them.

List of generic fault codes (DTCs)

SAE made up the original DTC list, but the manufacturers complained that they had their own systems and each system was different. Mercedes has a completely different system than Honda, etc. They cannot use each other's codes, so the SAE made a provision to separate the standard codes (P0 codes - see following table) from manufacturer specific codes (P1 codes).

P0100 - Mass or Volume Air Flow Circuit Malfunction
P0101 - Mass or Volume Air Flow Circuit Range/Performance Problem
P0102 - Mass or Volume Air Flow Circuit Low Input
P0103 - Mass or Volume Air Flow Circuit High Input
P0104 - Mass or Volume Air Flow Circuit Intermittent
P0105 - Manifold Absolute Pressure/Barometric Pressure Circuit Malfunction
P0106 - Manifold Absolute Pressure/Barometric Pressure Circuit Range/Performance Problem
P0107 - Manifold Absolute Pressure/Barometric Pressure Circuit Low Input
P0108 - Manifold Absolute Pressure/Barometric Pressure Circuit High Input
P0109 - Manifold Absolute Pressure/Barometric Pressure Circuit Intermittent
P0110 - Intake Air Temperature Circuit Malfunction
P0111 - Intake Air Temperature Circuit Range/Performance Problem
P0112 - Intake Air Temperature Circuit Low Input
P0113 - Intake Air Temperature Circuit High Input
P0114 - Intake Air Temperature Circuit Intermittent

P0115 - Engine Coolant Temperature Circuit Malfunction
P0116 - Engine Coolant Temperature Circuit Range/Performance Problem
P0117 - Engine Coolant Temperature Circuit Low Input
P0118 - Engine Coolant Temperature Circuit High Input
P0119 - Engine Coolant Temperature Circuit Intermittent
P0120 - Throttle/Petal Position Sensor/Switch A Circuit Malfunction
P0121 - Throttle/Petal Position Sensor/Switch A Circuit Range/Performance Problem
P0122 - Throttle/Petal Position Sensor/Switch A Circuit Low Input
P0123 - Throttle/Petal Position Sensor/Switch A Circuit High Input
P0124 - Throttle/Petal Position Sensor/Switch A Circuit Intermittent
P0125 - Insufficient Coolant Temperature for Closed Loop Fuel Control
P0126 - Insufficient Coolant Temperature for Stable Operation
P0130 - O2 Sensor Circuit Malfunction (Bank 1 Sensor 1)
P0131 - O2 Sensor Circuit Low Voltage (Bank 1 Sensor 1)
P0132 - O2 Sensor Circuit High Voltage (Bank 1 Sensor 1)
P0133 - O2 Sensor Circuit Slow Response (Bank 1 Sensor 1)
P0134 - O2 Sensor Circuit No Activity Detected (Bank 1 Sensor 1)
P0135 - O2 Sensor Heater Circuit Malfunction (Bank 1 Sensor 1)
P0136 - O2 Sensor Circuit Malfunction (Bank 1 Sensor 2)
P0137 - O2 Sensor Circuit Low Voltage (Bank 1 Sensor 2)
P0138 - O2 Sensor Circuit High Voltage (Bank 1 Sensor 2)
P0139 - O2 Sensor Circuit Slow Response (Bank 1 Sensor 2)
P0140 - O2 Sensor Circuit No Activity Detected (Bank 1 Sensor 2)
P0141 - O2 Sensor Heater Circuit Malfunction (Bank 1 Sensor 2)
P0142 - O2 Sensor Circuit Malfunction (Bank 1 Sensor 3)
P0143 - O2 Sensor Circuit Low Voltage (Bank 1 Sensor 3)
P0144 - O2 Sensor Circuit High Voltage (Bank 1 Sensor 3)
P0145 - O2 Sensor Circuit Slow Response (Bank 1 Sensor 3)
P0146 - O2 Sensor Circuit No Activity Detected (Bank 1 Sensor 3)
P0147 - O2 Sensor Heater Circuit Malfunction (Bank 1 Sensor 3)

P0150 - O2 Sensor Circuit Malfunction (Bank 2 Sensor 1)
P0151 - O2 Sensor Circuit Low Voltage (Bank 2 Sensor 1)
P0152 - O2 Sensor Circuit High Voltage (Bank 2 Sensor 1)
P0153 - O2 Sensor Circuit Slow Response (Bank 2 Sensor 1)
P0154 - O2 Sensor Circuit No Activity Detected (Bank 2 Sensor 1)
P0155 - O2 Sensor Heater Circuit Malfunction (Bank 2 Sensor 1)
P0156 - O2 Sensor Circuit Malfunction (Bank 2 Sensor 2)
P0157 - O2 Sensor Circuit Low Voltage (Bank 2 Sensor 2)
P0158 - O2 Sensor Circuit High Voltage (Bank 2 Sensor 2)
P0159 - O2 Sensor Circuit Slow Response (Bank 2 Sensor 2)
P0160 - O2 Sensor Circuit No Activity Detected (Bank 2 Sensor 2)
P0161 - O2 Sensor Heater Circuit Malfunction (Bank 2 Sensor 2)
P0162 - O2 Sensor Circuit Malfunction (Bank 2 Sensor 3)
P0163 - O2 Sensor Circuit Low Voltage (Bank 2 Sensor 3)
P0164 - O2 Sensor Circuit High Voltage (Bank 2 Sensor 3)
P0165 - O2 Sensor Circuit Slow Response (Bank 2 Sensor 3)
P0166 - O2 Sensor Circuit No Activity Detected (Bank 2 Sensor 3)
P0167 - O2 Sensor Heater Circuit Malfunction (Bank 2 Sensor 3)
P0170 - Fuel Trim Malfunction (Bank 1)
P0171 - System too Lean (Bank 1)
P0172 - System too Rich (Bank 1)
P0173 - Fuel Trim Malfunction (Bank 2)
P0174 - System too Lean (Bank 2)
P0175 - System too Rich (Bank 2)
P0176 - Fuel Composition Sensor Circuit Malfunction
P0177 - Fuel Composition Sensor Circuit Range/Performance
P0178 - Fuel Composition Sensor Circuit Low Input
P0179 - Fuel Composition Sensor Circuit High Input
P0180 - Fuel Temperature Sensor A Circuit Malfunction
P0181 - Fuel Temperature Sensor A Circuit Range/Performance
P0182 - Fuel Temperature Sensor A Circuit Low Input
P0183 - Fuel Temperature Sensor A Circuit High Input
P0184 - Fuel Temperature Sensor A Circuit Intermittent

Chapter 12 > OBDII diagnostic

P0185 - Fuel Temperature Sensor B Circuit Malfunction
P0186 - Fuel Temperature Sensor B Circuit Range/Performance
P0187 - Fuel Temperature Sensor B Circuit Low Input
P0188 - Fuel Temperature Sensor B Circuit High Input
P0189 - Fuel Temperature Sensor B Circuit Intermittent
P0190 - Fuel Rail Pressure Sensor Circuit Malfunction
P0191 - Fuel Rail Pressure Sensor Circuit Range/Performance
P0192 - Fuel Rail Pressure Sensor Circuit Low Input
P0193 - Fuel Rail Pressure Sensor Circuit High Input
P0194 - Fuel Rail Pressure Sensor Circuit Intermittent
P0195 - Engine Oil Temperature Sensor Malfunction
P0196 - Engine Oil Temperature Sensor Range/Performance
P0197 - Engine Oil Temperature Sensor Low
P0198 - Engine Oil Temperature Sensor High
P0199 - Engine Oil Temperature Sensor Intermittent
P0200 - Injector Circuit Malfunction
P0201 - Injector Circuit Malfunction - Cylinder 1
P0202 - Injector Circuit Malfunction - Cylinder 2
P0203 - Injector Circuit Malfunction - Cylinder 3
P0204 - Injector Circuit Malfunction - Cylinder 4
P0205 - Injector Circuit Malfunction - Cylinder 5
P0206 - Injector Circuit Malfunction - Cylinder 6
P0207 - Injector Circuit Malfunction - Cylinder 7
P0208 - Injector Circuit Malfunction - Cylinder 8
P0209 - Injector Circuit Malfunction - Cylinder 9
P0210 - Injector Circuit Malfunction - Cylinder 10
P0211 - Injector Circuit Malfunction - Cylinder 11
P0212 - Injector Circuit Malfunction - Cylinder 12
P0213 - Cold Start Injector 1 Malfunction
P0214 - Cold Start Injector 2 Malfunction
P0215 - Engine Shutoff Solenoid Malfunction
P0216 - Injection Timing Control Circuit Malfunction
P0217 - Engine Overtemp Condition
P0218 - Transmission Over Temperature Condition

P0219 - Engine Overspeed Condition
P0220 - Throttle/Petal Position Sensor/Switch B Circuit Malfunction
P0221 - Throttle/Petal Position Sensor/Switch B Circuit Range/Performance Problem
P0222 - Throttle/Petal Position Sensor/Switch B Circuit Low Input
P0223 - Throttle/Petal Position Sensor/Switch B Circuit High Input
P0224 - Throttle/Petal Position Sensor/Switch B Circuit Intermittent
P0225 - Throttle/Petal Position Sensor/Switch C Circuit Malfunction
P0226 - Throttle/Petal Position Sensor/Switch C Circuit Range/Performance Problem
P0227 - Throttle/Petal Position Sensor/Switch C Circuit Low Input
P0228 - Throttle/Petal Position Sensor/Switch C Circuit High Input
P0229 - Throttle/Petal Position Sensor/Switch C Circuit Intermittent
P0230 - Fuel Pump Primary Circuit Malfunction
P0231 - Fuel Pump Secondary Circuit Low
P0232 - Fuel Pump Secondary Circuit High
P0233 - Fuel Pump Secondary Circuit Intermittent
P0234 - Engine Overboost Condition
P0235 - Turbocharger Boost Sensor A Circuit Malfunction
P0236 - Turbocharger Boost Sensor A Circuit Range/Performance
P0237 - Turbocharger Boost Sensor A Circuit Low
P0238 - Turbocharger Boost Sensor A Circuit High
P0239 - Turbocharger Boost Sensor B Malfunction
P0240 - Turbocharger Boost Sensor B Circuit Range/Performance
P0241 - Turbocharger Boost Sensor B Circuit Low
P0242 - Turbocharger Boost Sensor B Circuit High
P0243 - Turbocharger Wastegate Solenoid A Malfunction
P0244 - Turbocharger Wastegate Solenoid A Range/Performance
P0245 - Turbocharger Wastegate Solenoid A Low
P0246 - Turbocharger Wastegate Solenoid A High
P0247 - Turbocharger Wastegate Solenoid B Malfunction
P0248 - Turbocharger Wastegate Solenoid B Range/Performance
P0249 - Turbocharger Wastegate Solenoid B Low
P0250 - Turbocharger Wastegate Solenoid B High

P0251 - Injection Pump Fuel Metering Control "A" Malfunction (Cam/Rotor/Injector)
P0252 - Injection Pump Fuel Metering Control "A" Range/Performance (Cam/Rotor/Injector)
P0253 - Injection Pump Fuel Metering Control "A" Low (Cam/Rotor/Injector)
P0254 - Injection Pump Fuel Metering Control "A" High (Cam/Rotor/Injector)
P0255 - Injection Pump Fuel Metering Control "A" Intermittent (Cam/Rotor/Injector)
P0256 - Injection Pump Fuel Metering Control "B" Malfunction (Cam/Rotor/Injector)
P0257 - Injection Pump Fuel Metering Control "B" Range/Performance (Cam/Rotor/Injector)
P0258 - Injection Pump Fuel Metering Control "B" Low (Cam/Rotor/Injector)
P0259 - Injection Pump Fuel Metering Control "B" High (Cam/Rotor/Injector)
P0260 - Injection Pump Fuel Metering Control "B" Intermittent (Cam/Rotor/Injector)
P0261 - Cylinder 1 Injector Circuit Low
P0262 - Cylinder 1 Injector Circuit High
P0263 - Cylinder 1 Contribution/Balance Fault
P0264 - Cylinder 2 Injector Circuit Low
P0265 - Cylinder 2 Injector Circuit High
P0266 - Cylinder 2 Contribution/Balance Fault
P0267 - Cylinder 3 Injector Circuit Low
P0268 - Cylinder 3 Injector Circuit High
P0269 - Cylinder 3 Contribution/Balance Fault
P0270 - Cylinder 4 Injector Circuit Low
P0271 - Cylinder 4 Injector Circuit High
P0272 - Cylinder 4 Contribution/Balance Fault
P0273 - Cylinder 5 Injector Circuit Low
P0274 - Cylinder 5 Injector Circuit High
P0275 - Cylinder 5 Contribution/Balance Fault
P0276 - Cylinder 6 Injector Circuit Low
P0277 - Cylinder 6 Injector Circuit High
P0278 - Cylinder 6 Contribution/Balance Fault
P0279 - Cylinder 7 Injector Circuit Low
P0280 - Cylinder 7 Injector Circuit High
P0281 - Cylinder 7 Contribution/Balance Fault
P0282 - Cylinder 8 Injector Circuit Low
P0283 - Cylinder 8 Injector Circuit High
P0284 - Cylinder 8 Contribution/Balance Fault
P0285 - Cylinder 9 Injector Circuit Low
P0286 - Cylinder 9 Injector Circuit High
P0287 - Cylinder 9 Contribution/Balance Fault
P0288 - Cylinder 10 Injector Circuit Low
P0289 - Cylinder 10 Injector Circuit High

P0290 - Cylinder 10 Contribution/Balance Fault
P0291 - Cylinder 11 Injector Circuit Low
P0292 - Cylinder 11 Injector Circuit High
P0293 - Cylinder 11 Contribution/Balance Fault
P0294 - Cylinder 12 Injector Circuit Low
P0295 - Cylinder 12 Injector Circuit High
P0296 - Cylinder 12 Contribution/Range Fault
P0300 - Random/Multiple Cylinder Misfire Detected
P0301 - Cylinder 1 Misfire Detected
P0302 - Cylinder 2 Misfire Detected
P0303 - Cylinder 3 Misfire Detected
P0304 - Cylinder 4 Misfire Detected
P0305 - Cylinder 5 Misfire Detected
P0306 - Cylinder 6 Misfire Detected
P0307 - Cylinder 7 Misfire Detected
P0308 - Cylinder 8 Misfire Detected
P0309 - Cylinder 9 Misfire Detected
P0311 - Cylinder 11 Misfire Detected
P0312 - Cylinder 12 Misfire Detected
P0320 - Ignition/Distributor Engine Speed Input Circuit Malfunction
P0321 - Ignition/Distributor Engine Speed Input Circuit Range/Performance
P0322 - Ignition/Distributor Engine Speed Input Circuit No Signal
P0323 - Ignition/Distributor Engine Speed Input Circuit Intermittent
P0325 - Knock Sensor 1 Circuit Malfunction (Bank 1 or Single Sensor)
P0326 - Knock Sensor 1 Circuit Range/Performance (Bank 1 or Single Sensor)
P0327 - Knock Sensor 1 Circuit Low Input (Bank 1 or Single Sensor)
P0328 - Knock Sensor 1 Circuit High Input (Bank 1 or Single Sensor)
P0329 - Knock Sensor 1 Circuit Intermittent (Bank 1 or Single Sensor)
P0330 - Knock Sensor 2 Circuit Malfunction (Bank 2)
P0331 - Knock Sensor 2 Circuit Range/Performance (Bank 2)
P0332 - Knock Sensor 2 Circuit Low Input (Bank 2)
P0333 - Knock Sensor 2 Circuit High Input (Bank 2)
P0334 - Knock Sensor 2 Circuit Intermittent (Bank 2)
P0335 - Crankshaft Position Sensor A Circuit Malfunction
P0336 - Crankshaft Position Sensor A Circuit Range/Performance
P0337 - Crankshaft Position Sensor A Circuit Low Input
P0338 - Crankshaft Position Sensor A Circuit High Input
P0339 - Crankshaft Position Sensor A Circuit Intermittent

P0340 - Camshaft Position Sensor Circuit Malfunction
P0341 - Camshaft Position Sensor Circuit Range/Performance
P0342 - Camshaft Position Sensor Circuit Low Input
P0343 - Camshaft Position Sensor Circuit High Input
P0344 - Camshaft Position Sensor Circuit Intermittent
P0350 - Ignition Coil Primary/Secondary Circuit Malfunction
P0351 - Ignition Coil A Primary/Secondary Circuit Malfunction
P0352 - Ignition Coil B Primary/Secondary Circuit Malfunction
P0353 - Ignition Coil C Primary/Secondary Circuit Malfunction
P0354 - Ignition Coil D Primary/Secondary Circuit Malfunction
P0355 - Ignition Coil E Primary/Secondary Circuit Malfunction
P0356 - Ignition Coil F Primary/Secondary Circuit Malfunction
P0357 - Ignition Coil G Primary/Secondary Circuit Malfunction
P0358 - Ignition Coil H Primary/Secondary Circuit Malfunction
P0359 - Ignition Coil I Primary/Secondary Circuit Malfunction
P0360 - Ignition Coil J Primary/Secondary Circuit Malfunction
P0361 - Ignition Coil K Primary/Secondary Circuit Malfunction
P0362 - Ignition Coil L Primary/Secondary Circuit Malfunction
P0370 - Timing Reference High Resolution Signal A Malfunction
P0371 - Timing Reference High Resolution Signal A Too Many Pulses
P0372 - Timing Reference High Resolution Signal A Too Few Pulses
P0373 - Timing Reference High Resolution Signal A Intermittent/Erratic Pulses
P0374 - Timing Reference High Resolution Signal A No Pulses
P0375 - Timing Reference High Resolution Signal B Malfunction
P0376 - Timing Reference High Resolution Signal B Too Many Pulses
P0377 - Timing Reference High Resolution Signal B Too Few Pulses
P0378 - Timing Reference High Resolution Signal B Intermittent/Erratic Pulses
P0379 - Timing Reference High Resolution Signal B No Pulses
P0380 - Glow Plug/Heater Circuit "A" Malfunction
P0381 - Glow Plug/Heater Indicator Circuit Malfunction

P0382 - Glow Plug/Heater Circuit "B" Malfunction
P0385 - Crankshaft Position Sensor B Circuit Malfunction
P0386 - Crankshaft Position Sensor B Circuit Range/Performance
P0387 - Crankshaft Position Sensor B Circuit Low Input
P0388 - Crankshaft Position Sensor B Circuit High Input
P0389 - Crankshaft Position Sensor B Circuit Intermittent
P0400 - Exhaust Gas Recirculation Flow Malfunction
P0401 - Exhaust Gas Recirculation Flow Insufficient Detected
P0402 - Exhaust Gas Recirculation Flow Excessive Detected
P0403 - Exhaust Gas Recirculation Circuit Malfunction
P0404 - Exhaust Gas Recirculation Circuit Range/Performance
P0405 - Exhaust Gas Recirculation Sensor A Circuit Low
P0406 - Exhaust Gas Recirculation Sensor A Circuit High
P0407 - Exhaust Gas Recirculation Sensor B Circuit Low
P0408 - Exhaust Gas Recirculation Sensor B Circuit High
P0410 - Secondary Air Injection System Malfunction
P0411 - Secondary Air Injection System Incorrect Flow Detected
P0412 - Secondary Air Injection System Switching Valve A Circuit Malfunction
P0413 - Secondary Air Injection System Switching Valve A Circuit Open
P0414 - Secondary Air Injection System Switching Valve A Circuit Shorted
P0415 - Secondary Air Injection System Switching Valve B Circuit Malfunction
P0416 - Secondary Air Injection System Switching Valve B Circuit Open
P0417 - Secondary Air Injection System Switching Valve B Circuit Shorted
P0418 - Secondary Air Injection System Relay "A" Circuit Malfunction
P0419 - Secondary Air Injection System Relay "B" Circuit Malfunction
P0420 - Catalyst System Efficiency Below Threshold (Bank 1)
P0421 - Warm Up Catalyst Efficiency Below Threshold (Bank 1)
P0422 - Main Catalyst Efficiency Below Threshold (Bank 1)
P0423 - Heated Catalyst Efficiency Below Threshold (Bank 1)
P0424 - Heated Catalyst Temperature Below Threshold (Bank 1)

P0430 - Catalyst System Efficiency Below Threshold (Bank 2)
P0431 - Warm Up Catalyst Efficiency Below Threshold (Bank 2)
P0432 - Main Catalyst Efficiency Below Threshold (Bank 2)
P0433 - Heated Catalyst Efficiency Below Threshold (Bank 2)
P0434 - Heated Catalyst Temperature Below Threshold (Bank 2)
P0440 - Evaporative Emission Control System Malfunction
P0441 - Evaporative Emission Control System Incorrect Purge Flow
P0442 - Evaporative Emission Control System Leak Detected (small leak)
P0443 - Evaporative Emission Control System Purge Control Valve Circuit Malfunction
P0444 - Evaporative Emission Control System Purge Control Valve Circuit Open
P0445 - Evaporative Emission Control System Purge Control Valve Circuit Shorted
P0446 - Evaporative Emission Control System Vent Control Circuit Malfunction
P0447 - Evaporative Emission Control System Vent Control Circuit Open
P0448 - Evaporative Emission Control System Vent Control Circuit Shorted
P0449 - Evaporative Emission Control System Vent Valve/Solenoid Circuit Malfunction
P0450 - Evaporative Emission Control System Pressure Sensor Malfunction
P0451 - Evaporative Emission Control System Pressure Sensor Range/Performance
P0452 - Evaporative Emission Control System Pressure Sensor Low Input
P0453 - Evaporative Emission Control System Pressure Sensor High Input
P0454 - Evaporative Emission Control System Pressure Sensor Intermittent
P0455 - Evaporative Emission Control System Leak Detected (gross leak)
P0460 - Fuel Level Sensor Circuit Malfunction
P0461 - Fuel Level Sensor Circuit Range/Performance
P0462 - Fuel Level Sensor Circuit Low Input
P0463 - Fuel Level Sensor Circuit High Input
P0464 - Fuel Level Sensor Circuit Intermittent
P0465 - Purge Flow Sensor Circuit Malfunction
P0466 - Purge Flow Sensor Circuit Range/Performance
P0467 - Purge Flow Sensor Circuit Low Input
P0468 - Purge Flow Sensor Circuit High Input
P0469 - Purge Flow Sensor Circuit Intermittent
P0470 - Exhaust Pressure Sensor Malfunction
P0471 - Exhaust Pressure Sensor Range/Performance
P0472 - Exhaust Pressure Sensor Low
P0473 - Exhaust Pressure Sensor High

P0474 - Exhaust Pressure Sensor Intermittent
P0475 - Exhaust Pressure Control Valve Malfunction
P0476 - Exhaust Pressure Control Valve Range/Performance
P0477 - Exhaust Pressure Control Valve Low
P0478 - Exhaust Pressure Control Valve High
P0479 - Exhaust Pressure Control Valve Intermittent
P0480 - Cooling Fan 1 Control Circuit Malfunction
P0481 - Cooling Fan 2 Control Circuit Malfunction
P0482 - Cooling Fan 3 Control Circuit Malfunction
P0483 - Cooling Fan Rationality Check Malfunction
P0484 - Cooling Fan Circuit Over Current
P0485 - Cooling Fan Power/Ground Circuit Malfunction
P0500 - Vehicle Speed Sensor Malfunction
P0501 - Vehicle Speed Sensor Range/Performance
P0502 - Vehicle Speed Sensor Low Input
P0503 - Vehicle Speed Sensor Intermittent/Erratic/High
P0505 - Idle Control System Malfunction
P0506 - Idle Control System RPM Lower Than Expected
P0507 - Idle Control System RPM Higher Than Expected
P0510 - Closed Throttle Position Switch Malfunction
P0520 - Engine Oil Pressure Sensor/Switch Circuit Malfunction
P0521 - Engine Oil Pressure Sensor/Switch Circuit Range/Performance
P0522 - Engine Oil Pressure Sensor/Switch Circuit Low Voltage
P0523 - Engine Oil Pressure Sensor/Switch Circuit High Voltage
P0530 - A/C Refrigerant Pressure Sensor Circuit Malfunction
P0531 - A/C Refrigerant Pressure Sensor Circuit Range/Performance
P0532 - A/C Refrigerant Pressure Sensor Circuit Low Input
P0533 - A/C Refrigerant Pressure Sensor Circuit High Input
P0534 - Air Conditioner Refrigerant Charge Loss
P0550 - Power Steering Pressure Sensor Circuit Malfunction
P0551 - Power Steering Pressure Sensor Circuit Range/Performance
P0552 - Power Steering Pressure Sensor Circuit Low Input
P0553 - Power Steering Pressure Sensor Circuit High Input
P0554 - Power Steering Pressure Sensor Circuit Intermittent

P0560 - System Voltage Malfunction
P0561 - System Voltage Unstable
P0562 - System Voltage Low
P0563 - System Voltage High
P0565 - Cruise Control On Signal Malfunction
P0566 - Cruise Control Off Signal Malfunction
P0567 - Cruise Control Resume Signal Malfunction
P0568 - Cruise Control Set Signal Malfunction
P0569 - Cruise Control Coast Signal Malfunction
P0570 - Cruise Control Accel Signal Malfunction
P0571 - Cruise Control/Brake Switch A Circuit Malfunction
P0572 - Cruise Control/Brake Switch A Circuit Low
P0573 - Cruise Control/Brake Switch A Circuit High
P0574 - Cruise Control Related Malfunction
P0575 - Cruise Control Related Malfunction
P0576 - Cruise Control Related Malfunction
P0577 - Cruise Control Related Malfunction
P0578 - Cruise Control Related Malfunction
P0579 - Cruise Control Related Malfunction
P0580 - Cruise Control Related Malfunction
P0600 - Serial Communication Link Malfunction
P0601 - Internal Control Module Memory Check Sum Error
P0602 - Control Module Programming Error
P0603 - Internal Control Module Keep Alive Memory (KAM) Error
P0604 - Internal Control Module Random Access Memory (RAM) Error
P0605 - Internal Control Module Read Only Memory (ROM) Error
P0606 - PCM Processor Fault
P0608 - Control Module VSS Output "A" Malfunction
P0609 - Control Module VSS Output "B" Malfunction
P0620 - Generator Control Circuit Malfunction
P0621 - Generator Lamp "L" Control Circuit Malfunction
P0622 - Generator Field "F" Control Circuit Malfunction
P0650 - Malfunction Indicator Lamp (MIL) Control Circuit Malfunction
P0654 - Engine RPM Output Circuit Malfunction
P0655 - Engine Hot Lamp Output Control Circuit Malfucntion
P0656 - Fuel Level Output Circuit Malfunction
P0700 - Transmission Control System Malfunction

P0701 - Transmission Control System Range/Performance
P0702 - Transmission Control System Electrical
P0703 - Torque Converter/Brake Switch B Circuit Malfunction
P0704 - Clutch Switch Input Circuit Malfunction
P0705 - Transmission Range Sensor Circuit malfunction (PRNDL Input)
P0706 - Transmission Range Sensor Circuit Range/Performance
P0707 - Transmission Range Sensor Circuit Low Input
P0708 - Transmission Range Sensor Circuit High Input
P0709 - Transmission Range Sensor Circuit Intermittent
P0710 - Transmission Fluid Temperature Sensor Circuit Malfunction
P0711 - Transmission Fluid Temperature Sensor Circuit Range/Performance
P0712 - Transmission Fluid Temperature Sensor Circuit Low Input
P0713 - Transmission Fluid Temperature Sensor Circuit High Input
P0714 - Transmission Fluid Temperature Sensor Circuit Intermittent
P0715 - Input/Turbine Speed Sensor Circuit Malfunction
P0716 - Input/Turbine Speed Sensor Circuit Range/Performance
P0717 - Input/Turbine Speed Sensor Circuit No Signal
P0718 - Input/Turbine Speed Sensor Circuit Intermittent
P0719 - Torque Converter/Brake Switch B Circuit Low
P0720 - Output Speed Sensor Circuit Malfunction
P0721 - Output Speed Sensor Range/Performance
P0722 - Output Speed Sensor No Signal
P0723 - Output Speed Sensor Intermittent
P0724 - Torque Converter/Brake Switch B Circuit High
P0725 - Engine Speed input Circuit Malfunction
P0726 - Engine Speed Input Circuit Range/Performance
P0727 - Engine Speed Input Circuit No Signal
P0728 - Engine Speed Input Circuit Intermittent
P0730 - Incorrect Gear Ratio
P0731 - Gear 1 Incorrect ratio
P0732 - Gear 2 Incorrect ratio
P0733 - Gear 3 Incorrect ratio
P0734 - Gear 4 Incorrect ratio
P0735 - Gear 5 Incorrect ratio
P0736 - Reverse incorrect gear ratio

P0740 - Torque Converter Clutch Circuit Malfunction
P0741 - Torque Converter Clutch Circuit Performance or Stuck Off
P0742 - Torque Converter Clutch Circuit Stuck On
P0743 - Torque Converter Clutch Circuit Electrical
P0744 - Torque Converter Clutch Circuit Intermittent
P0745 - Pressure Control Solenoid Malfunction
P0746 - Pressure Control Solenoid Performance or Stuck Off
P0747 - Pressure Control Solenoid Stuck On
P0748 - Pressure Control Solenoid Electrical
P0749 - Pressure Control Solenoid Intermittent
P0750 - Shift Solenoid A Malfunction
P0751 - Shift Solenoid A Performance or Stuck Off
P0752 - Shift Solenoid A Stuck On
P0753 - Shift Solenoid A Electrical
P0754 - Shift Solenoid A Intermittent
P0755 - Shift Solenoid B Malfunction
P0756 - Shift Solenoid B Performance or Stuck Off
P0757 - Shift Solenoid B Stuck On
P0758 - Shift Solenoid B Electrical
P0759 - Shift Solenoid B Intermittent
P0760 - Shift Solenoid C Malfunction
P0761 - Shift Solenoid C Performance or Stuck Off
P0762 - Shift Solenoid C Stuck On
P0763 - Shift Solenoid C Electrical
P0764 - Shift Solenoid C Intermittent
P0765 - Shift Solenoid D Malfunction
P0766 - Shift Solenoid D Performance or Stuck Off
P0767 - Shift Solenoid D Stuck On
P0768 - Shift Solenoid D Electrical
P0769 - Shift Solenoid D Intermittent
P0770 - Shift Solenoid E Malfunction
P0771 - Shift Solenoid E Performance or Stuck Off
P0772 - Shift Solenoid E Stuck On
P0773 - Shift Solenoid E Electrical
P0774 - Shift Solenoid E Intermittent
P0780 - Shift Malfunction
P0781 - 1-2 Shift Malfunction
P0782 - 2-3 Shift Malfunction
P0783 - 3-4 Shift Malfunction
P0784 - 4-5 Shift Malfunction
P0785 - Shift/Timing Solenoid Malfunction

P0786 - Shift/Timing Solenoid Range/Performance
P0787 - Shift/Timing Solenoid Low
P0788 - Shift/Timing Solenoid High
P0789 - Shift/Timing Solenoid Intermittent
P0790 - Normal/Performance Switch Circuit Malfunction
P0801 - Reverse Inhibit Control Circuit Malfunction
P0803 - 1-4 Upshift (Skip Shift) Solenoid Control Circuit Malfunction
P0804 - 1-4 Upshift (Skip Shift) Lamp Control Circuit Malfunction

Standardized component and emissions terminology

As in any trade, there are different names and slang for the same word. Take fault code for instance. Some call it a code, others call it a fault, and still others call it "that thang that's broke". The EPA came up with a standardized list of terminology for the components that pertain to OBDII. A DTC is a fault, code, or "that thang that's broke". As always when a governmental organization is involved, acronyms, and jargon is a must.

This is a basic list of the acronyms:

AFC - Air Flow Control
CAN - Controller Area Network
CARB - California Air Resources Board
CFI - Continuous Fuel Injection
CO - Carbon Monoxide
DLC - Data Link Connector
DTC - Diagnostic Trouble Code
ECM - Engine Control Module- usually the main in-car computer controlling emissions and engine operation
ECT - Engine Coolant Temperature
EEC - Electronic Engine Control
EEPROM or **E²PROM** - Electrically Erasable Programmable Read Only Memory

Chapter 12 > OBDII diagnostic

EFI - Electronic Fuel Injection
EGR - Exhaust Gas Recirculation
EMR - Electronic Module Retard
ESC - Electronic Spark Control
EST - Electronic Spark Timing
DTC - Diagnostic Trouble Code
FLI - Fuel Level Indicator
HC - Hydrocarbons
HEI - High Energy Ignition
HO$_2$S - Heated Oxygen Sensor
MAF - Mass Air Flow
MAP - Manifold Absolute Pressure
MAT - Manifold Air Temperature
MFG - Manufacturer
MIL - Malfunction Indicator Light. The "Check Engine Light" on your dash.
NOx - Oxides of Nitrogen
O$_2$ - Oxygen
OBD - On-Board Diagnostics
PCM - Powertrain Control Module, the on-board computer that controls engine and drive train
PCV - Positive Crankcase Ventilation
PID - Parameter ID
PTC - Pending Trouble Code
RPM - Revolutions Per Minute
SES - Service Engine Soon dash light, now referred to as MIL
SFI - Sequential Fuel Injection
Stoichiometric (Stoy'-kee-o-metric)
Ratio - Theoretical perfect combustion ratio of 1 part gas to 14.7 parts air
TBI - Throttle Body Injection
TPI - Tuned Port Injection
TPS - Throttle Position Sensor
VAC - Vacuum
VCM - Vehicle Control Module, the in-car computer that oversees engine management, transmission operation, anti-lock brakes, and other functions not directly related to emissions control
VIN - Vehicle Identification Number
VSS - Vehicle Speed Sensor
WOT - Wide Open Throttle

Conclusion

Today the technician cannot just listen to an engine and repair it, as in the past. If we were to remove a spark plug, the engine management will adjust to that missing cylinder. Diagnosis has become much more complex, because manufacturers had to design a vehicle complying with the standards of OBDII. In other words, in order to sell in the United States, manufacturers had to design a new, more complex vehicle around OBDII.

SECRETS REVEALED

In the past, manufacturers basically added a few parts to an existing vehicle to comply with emissions. If an emission related part failed, the mechanic would just remove it. The vehicle would run fine, for it was not originally designed for the part to begin with.

Today the vehicles are made for OBDII. That means they are made to work with this system. If a mechanic were to remove defective parts the vehicle would not run properly. The mechanic of yesterday MUST become a technician today. A mechanic will have knowledge of how to repair a vehicle, and usually a great knowledge. The technician of today must posses the knowledge of computer science plus mechanical knowledge. The technician must know how to read waveforms from an oscilloscope; how data buses work; how to read complex wiring diagrams, etc. OBDII is responsible for this movement in technology.

With the complexity of the vehicles of today, the use of a scan tool is now mandatory. Not all scan tools are created equal. Before the technician invests in a scan tool, he must look at what the tool will do.

A good all-around scan tool must be able to:

- Give a complete list of the systems monitored, their status, and continuous misfire detection.

- Retrieve faults (DTC's), both stored, and active. (To be a complete tool, it should also show the manufacturer enhanced P1 codes, the description and be able to reset the DTC's).

- Show "Freeze Frame" status of the DTC (the snapshot of the situation when the DTC occurred).

- Show the status of the Oxygen sensors.

- Show a complete list of the "Actual Values" (the real time data values of the engine components).

As you can see, there are Millions of vehicles on the road with OBDII. Like it or not, OBDII has influenced your life, your children's lives and, hopefully their children's lives with cleaner air and a cleaner environment.

This chapter will take you through one complete run in a Wild Street low – 8 – second car. Willie will take you through inspection, staging, and crossing the 1/4 mile mark.

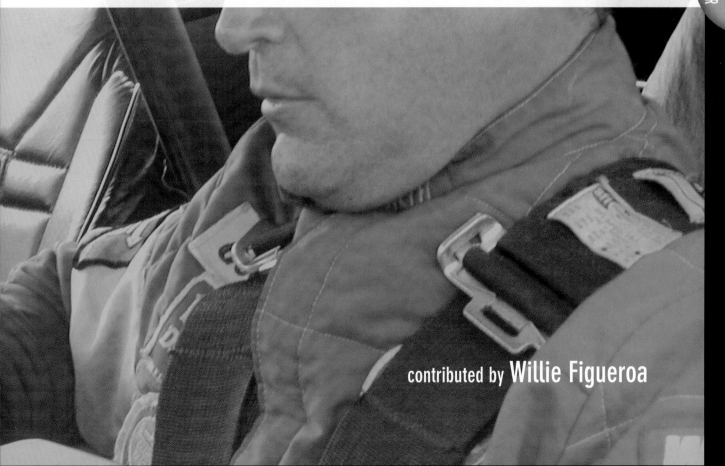

Chapter 13 > 1500hp down the 1/4 Mile with Willie Figueroa

contributed by Willie Figueroa

Chapter 13 > 1500hp down the 1/4 Mile with Willie Figueroa

FORD TUNING

There are many critical aspects involved in drag racing. Many do's and don'ts. Making mistakes at these velocities could be potentially fatal, therefore it is very important to follow a strict schedule when preparing and racing these vehicles.

First, I inspect the track for imperfections such as bumps and cracks. It is important to find the "groove". The groove is created from the indentations of the vehicle tires when racing down the drag strip. By inspecting the track surface and finding the racing groove, one can better select and prepare for lane choice. I also observe the fast cars and their favorite lane preferences. Always select the lane with the fewest blemishes and with the most obvious groove. During qualifying one is assigned to a different lane each run.

During rounds of elimination the fast cars have lane preference. I try to stay in the same lane throughout the rounds.

Second, the vehicle (my most recent experience was a supercharged SN95 body with a 370inch Windsor motor owned by UPR Racing) must be inspected. My crew and I always ensure the car is race ready an hour before the round. We check the fuel level and make sure the tire pressure is at my preset preference of 13psi. I also double-check everything under the hood previously checked by a crewmember. Items that I inspect are spark plug wires, radiator cap, hoses and clamps (especially on the intake side from the supercharger which is under heavy pressure). The vehicle is always under constant supervision by the crew. Last, but not least, we make sure there is a white shoe polish smiley face on the back window.

Third, once called to the lanes, my crewmembers or I will fill the intercooler tank with ice. After that I jump into the car, bring the 1300+ hp Windsor to life and drive to the lane. I make sure to obey all the traffic

SECRETS REVEALED

Vibeke (who has been a trusted crew member for the last three wining years) checks the tire pressure one last time, making sure that pressures are even. If uneven, it will affect the steering behavior of the car down the drag strip. Then, I drive into the water box, looking for the best position, which is usually at the edge of the water (this is the burnout area, to heat up your tires for traction).

Before I begin the burnout, I observe the car in front of me to see if everything went well with his run.

When I get the signal from my crew and track members to begin the burnout, I ease my foot off the brake, push the throttle, causing the tires to begin to spin and heat up. (**Note:** Willie's preference is a 5500rpm high gear burnout due to the fact that one would want to keep the engine rpm's low with a high rate of speed on the tires)

When my crew signals me to pull forward, I keep the engine at a constant rpm, usually through the starting line and 100-200 feet down the track. I enjoy a long burnout to put on a show for the racing fans. Who doesn't?

After the burnout, I try to back up on the rubber I just laid on the track, with the assistance of my crew.

Behind the starting line, I turn on the switch for the water intercooler pump and set the laptop to begin recording the run. The laptop is preset to begin recording at the start of the run. At that time I get a "thumbs up" from my crew, ensuring me the car is ready for flight.

I tighten my shoulder belts as tight as possible and continue to move to the pre-stage beam. Out of the corner of my eye, I observe the competitor in the other lane to ensure that we have similar staging times.

Once the pre-beam has been set (first light bulb on the Xmas tree), I concentrate entirely on the tree. At this time, I will see whether the other car has pre-staged or not and I will give staging lane courtesy. I have good staging ethics and will not burn anybody.

With both cars pre-staged at time, I always make sure I have the car in first gear. Then, with my foot heavily on the brakes, I will bring the engine 2,000rpm and ease off the foot brake until I move into the second staging beam. I follow by applying the trans brake button (if all staging lights are full). At that moment I enter W.O.T. (Wide Open Throttle), which is equal to about 4.65 Volts on the TPS (Throttle Position Sensor). Once the pro tree has been activated, I release the trans brake button and launch. I do the best to keep the car in the groove, if the tires break loose, I try to pedal it (lift off the gas a little). If that maneuver does not help, I proceed to void the run by returning the TPS to 0.98Volts (it is very important to realize one's limitations in driving).

Besides, you don't want to scratch your paint on the wall and wake up the emergency crew!

If I am in a good run and the engine reaches its shift point (8000rpm on UPR's power plant), I will move my right hand slightly and push the shifter into high gear (warp speed). At that time the engine should be at a high pitch and at about 30psi of boost.

With total focus on the finish line, I drop the laundry (parachute) 50ft before reaching the finish line. At this speed, I will pass the finish line before the parachute fully deploys to slow the car down. When I pass through the finish line, I put the car into neutral. Once the car has reached a safe speed, I take my eyes off the road, look at the gauges and proceed to the return lane. Once off the track, I stop the car and jump out to pick up the chute. After I get back into the car, I stop the recording and save the run on the laptop - I use it as reference for future tuning.

I continue to follow the direction of the track crew (which could include fuel check and weigh in) and perhaps share some thoughts of the run with some of the other drivers in the waiting lanes.

Once signaled by the track crew to continue, I proceed to the ticket booth and grab the ticket from my run, hoping it was a good one.

Lastly I continue to the pits to meet with my crew and discuss and view the run on the laptop; perhaps making some changes for the next round.

I want to say that winning a race involves more than just good driving skills. It takes a lot of hard work, practice, a good car, a good crew and a lot of luck. I am blessed to have all of these in my life.

Read about the principal
distributor of DiabloSport.

Chapter 14 > The PDQ Story

Chapter 14 > The PDQ Story

FORD TUNING

As one of the country's largest and best-known distribution and marketing firms of automotive performance products, PDQ Performance Warehouse has accomplished the American dream.

Born and raised in Yugoslavia, brothers Alex and Sam Nikolic spent most of their adolescence working and helping their parents run their family owned and operated businesses. They worked every evening after finishing their days at school.

In 1980, Yugoslavian President Josip "Tito" Broz passed away and the country's downward spiral began. Despite the country's unsecured state, life went on. Alex and Sam's real passion was automotive and mechanical engines. They sought ways of producing more efficient power for automotive and farm equipment.

In 1990, Sam traveled to the United States to attend pilot school and obtain a commercial pilot license - Alex had already obtained his in Yugoslavia.

In 1995 Alex joined Sam in the United States. After many months of research and market testing, PDQ Performance Warehouse was created. Their idea was to provide unique and technologically advanced products to help improve engine efficiency and power.

In order to be on the "cutting edge" of performance, PDQ conducted their own research and development and brought new technologies to the forefront.

During this time, Alex and Sam met Chris Stajdel (brother of famous tuner Patrick Stajdel). Born in Miami, FL, Chris has an engine performance background that dates back to 1977. An avid motorcycle racer, Chris worked on motorcycle engines to improve performance.

From the mid 1980's through 1990 Chris was involved with a nationally known high performance automotive shop that specialized in electronically controlled turbocharged vehicles. In just a few short years, the company exploded onto the performance scene and was recognized by many automotive magazines and enthusiasts. One of the company goals was to develop the quickest full-bodied, street driven (no wheel tubs) vehicle in the country. In 1989, their goal was accomplished. At over 3400 lbs., their full bodied 900+ horsepower twin turbo Buick Regal ran the quarter mile in 10.02 seconds at over 139 mph setting an unofficial national record.

In the year 2000, Chris joined PDQ, which continues to stay on the cutting edge of high performance specialty products. Current products include power enhancements for John Deere tractors, unique exhaust systems for heavy-duty equipment, transmission and engine monitoring gauges and state-of-the-art air induction kits.

One of their most impressive ventures is their strong relationship with DiabloSport. DiabloSport has already grown to be one of the country's most advanced companies involved in research, development and manufacturing of performance electronics.

Chris, Alex and Sam have set the company goals high. With their strong backgrounds and determination, we're confident they will achieve them.

If you have any interest in DiabloSport products or would like to become a dealer, contact Chris, Alex or Sam at:

PDQ Performance Whse.
PDQ Marketing Group
1470 Tropic Park Dr.
Sanford, FL 32773
407-321-5644
www.pdqperformance.com

Cooking with DiabloSport, hot and spicy

After a successful day of tuning, read the most important chapter! Treat your friends and family to a full course dinner; this time Hungarian style.

As you buy each of the Secrets Revealed books, just cut out the recipes, collect them, and you will have…well, just that, a collection.

ENTER

Life should be full of pleasures, and eating well is certainly one of them. Since you love performance, winning and winners, we are presenting a sure winner! When you prepare the following menu, everyone invited will be satisfied!

Afterwards, let us know, email us at fordtuning@kotzigpublishing.com. And, if you want to share your favorite winner menu with others, email it to the same address. You might want to include a paragraph or two about yourself, picture of your car, and whatever else you deem important. We might post it on the www.diablosport.com website, or even publish it in the next version of this book.

Hungarian menu – a sure winner!

Appetizer:

Bread with Farmer's cheese spread

[Körözött juhtúró kenyérrel]

First Course:

Veal filled pancakes from the Hortobadji region

[Hortobágyi palacsinta]

Second Course:

Pork and rice – filled green pepper in tomato sauce

[Töltött paprika]

Dessert

Farmer's cheese tart à la chef Rákóczi

[Rákóczi túrós lepény]

All the above dishes are selected with one thing in mind – we want the cook to enjoy the party as well! Everything can be cooked days in advance, and heated just before the frogs in your guests' bellies start to croak (as they say in Hungary).

Farmer's cheese spread

First let's make real Farmer's cheese:

(Non-purists can skip this paragraph and buy it ready - ask for 4 packages of unsalted farmer's cheese by Friendship at your local supermarket)

- *2 gallons of 2 - 4% milk (not ultra-pasteurized - use ordinary milk in a plastic bottle)*

- *1 cup of plain yoghurt (try to use one made from milk and active cultures only; no gelatin, wax or other additives)*

It might sound scary, to do something from scratch like our grandmothers used to do, but rest assured you will spend only minutes in the kitchen. Nature will do the rest.

Heat the milk until very warm (that is until you can still dip your finger into it for 5 seconds without burning yourself). Pour it into a glass or plastic container. Mix in the yoghurt and cover it with a plate or plastic-wrap.

Keep the mixture at room temperature for about 6 hours. Active cultures will convert the milk into a delicious and light yoghurt.

Note: Use the above procedure to make yoghurt for the whole family. You won't have to buy yoghurt ever again! It stays fresh in the refrigerator for days. Just take some of the previously made yoghurt and mix it with warm milk.
If you like your yoghurt firmer and sourer – just extend the 6 hours standing time to up to 24 hours.

Pour your homemade yoghurt into a pot and start heating. Stir slowly with a wooden spoon. Within minutes it will start to curdle and separate into a white mass and a greenish liquid. Lower the heat at this point. One minute later, when all the yoghurt has clearly separated, turn the heat off. The curdling process doesn't require high temperatures.

Pour the cheese mixture through a strainer. Leave it in the strainer for about 5 minutes.

Voilá! You have just made your own Farmer's cheese! Isn't it delicious (and easy)?

Put half of the farmer's cheese in the refrigerator – we will use it later for dessert preparation.

There are many uses for Farmer's cheese. It will refrigerate for many days and freeze for months.

COOKING WITH DIABLOSPORT
HUNGARIAN COOKING

Next, we will make an excellent spread from it. We guarantee that your guests will return again and again.

Now let's finish the spread:

- *1/2 amount of the above farmer's cheese*
- *1 stick of real butter, well softened at room temperature*
- *1 bunch of chives or a small onion, minced*
- *2 tablespoons of caraway seeds, crushed*
- *2 teaspoons of sweet red paprika (Hungarian if available)*
- *2 teaspoons of hot red paprika (Hungarian if available)*
- *1 teaspoon of ground pepper*
- *Salt to taste*

Mix all ingredients very well, until creamy.

This spread can stay in the refrigerator for weeks, or it can be frozen.

Make sure to serve it in spreadable state – same as butter, softened tastes so much better!

Veal filled crêpes (pancakes)
First, let's make 10 crêpes

- *2 eggs*
- *1 teaspoon of sugar*
- *1 teaspoon of salt*
- *2 cups of unbleached flour*
- *1 cup of water*
- *1 cup of milk*
- *olive oil for the pan*
- *A crêpe pan and a wooden spatula*

Put all ingredients in the blender and run on high for 1 minute. The dough must have the proper consistency for crêpes – it must run like honey, not like water. Let the dough rest for at least 15 minutes. If necessary, correct the consistency by adding water or flour.

Heat a good crêpe pan (if non-stick, you need much less oil). Brush it with a very small amount of oil. The crêpe is going to be dried on the pan; not fried in heavy oil. Leave the flame on medium heat.

Lift the pan from the stovetop and, using the other hand, pour the dough straight from the blender on 1/3 of the pan. Immediately tilt the pan in a circular motion to cover the bottom of the pan evenly. The proper thickness of the crêpe is about 1/10 of an inch.

Place the pan back on the stovetop. In less than a minute, the edges of the crêpe start to be visibly dried out; perhaps even start to get a little brown. Grab the pan and using the wooden spatula in your other hand, gently lift the edges of the pancake all around the pan. You should then be able to slide the spatula under the pancake and to turn it on the other side with a quick motion. Some brown spots on the crêpe are all right (the French wouldn't allow it – but this is better). The second side needs only 15 – 30 seconds to be done. Slide the crêpe out onto a large plate, where you will collect them one on top of another, until all the dough is used up.

The first crêpe usually doesn't work out – the dough is either too thick (doesn't cover the whole pan when tilting) or too thin (crêpe tears when turning). This can be solved by adding water or flour to the dough and blending it for another 30 seconds.

Crêpes can also be prepared in advance, as they stay fresh in the refrigerator for a few days when covered with plastic-wrap.

HUNGARIAN COOKING

Now let's make the veal filling (you can replace the veal with chicken if you want)

- *1 large onion, minced*
- *1/4 cup of extra virgin olive oil*
- *2 teaspoons of sweet red paprika (Hungarian if available)*
- *1 teaspoon of hot red paprika (Hungarian if available)*
- *1 cup of Marsala wine*
- *1 pound of veal, cut in thin strips*
- *1 tomato, cubed*
- *2 cups of Portobello mushrooms, cut to small pieces*
- *1 cup of sour cream*
- *1 teaspoon of ground pepper*
- *Salt to taste*

Liquefy the mushrooms, 1/2 of the wine and 1 cup of water in the blender. This will be our sauce when serving.

Put oil and the onion in a medium size pot. Heat until the onion is translucent – don't let it brown.

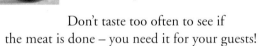

Pour all the paprika powder on the hot onion and mix with a wooden spoon, for no more than 15 seconds. Be careful not to burn the paprika powder.

Pour the remainings wine on the onion mixture to prevent the paprika from burning. Add all the veal. Stir occasionally, until all the veal is white.

Leave the veal mixture until all the liquid evaporates. Add all the remaining ingredients, cover and simmer for 20 minutes or until the meat is tender. If too dry, add water to keep it moist.

Don't taste too often to see if the meat is done – you need it for your guests!

Now let's finish it

- *Above crêpes*
- *Above veal filling*
- *1 cup of real sour cream*
- *1/4 cup of extra virgin olive oil*
- *2 eggs*
- *cup of water*
- *Covered casserole*

Preheat the oven to 375°F.

Fill each crêpe with veal filling and roll it. Place in the casserole, next to each other.

Mix all the sour cream, oil, eggs and water. Pour the mixture over the pancakes.

Baked covered for 30 minutes.

Place one crêpe on each plate and cover with hot mushroom sauce. Serve right away, or reheat later.

Pork and rice – filled green pepper in tomato sauce
First let's make the pork filling

- 1 large red onion, chopped
- 1/4 cup of olive oil
- 2 lbs of ground pork; or combination of pork and veal
- 1 teaspoon of salt
- 1 teaspoon of ground pepper
- 1 cup of rice (best is Italian risotto rice, Carnaroli or Arborio)
- 1 cup of water
- 1 egg
- 8 green peppers (Cubanelle, if available)

Sauté the onion in olive oil until translucent. Stir in the pork, salt and the black pepper. Add a little water, if necessary. Cook until done, about 20 minutes.

In another pot, mix rice and water and cook on low heat, until all the water is absorbed. Cook only until firm, we will continue cooking it later in the tomato sauce.

Let both the pork and rice cool so they are easier to handle. Mix together the pork, rice and the egg.

Cut off the stem end of the green peppers and remove the seeds. Fill the peppers with the meat mixture. Pack the filling well, so there are no air pockets. Any remaining meat mixture will be added to the tomato sauce.

Did you know that the word "paprika" is Hungarian and means pepper?

Now let's make the tomato sauce

- *4 large tomatoes, crashed or cubed*
- *1 large can (600g) of tomato paste*
- *1 large can (600g) of tomato sauce*
- *cup of olive oil*
- *6 cloves of garlic, mashed*

- *2 teaspoons of sweet red paprika (Hungarian if available)*
- *1 teaspoon of hot red paprika (Hungarian if available)*
- *2 cups of white wine*
- *8 cups of water*
- *Salt to taste*

Mix all the ingredients in a large pot. If there was any leftover meat mixture, add it to the sauce.

Now, let's finish the meal

Place all the green peppers into the pot with the tomato sauce, open end, up. If there isn't enough tomato sauce to cover the peppers, add water to the sauce.

Cover the pot and cook on a low heat for 30 minutes.

This meal can stay in the refrigerator for days, or it can be frozen. Just reheat before serving. Arrange one green pepper per serving on a deep plate and pour a generous amount of tomato sauce around it. Serve with a steak knife, fork and tablespoon.

Important note: hide your nicest white tablecloth in a closet!

Farmer's cheese tart

- *1 cup of flour*
- *1 stick of real butter, unsalted, softened(!)*
- *1 cup of sugar*
- *1 lemon*
- *1/3 cup of sour cream*
- *1 level tablespoon of baking powder*
- *3 eggs, separated*
- *1 additional egg yolk*
- *1 cup of jam (try red current or apricot)*
- *2 cups of farmer's cheese (see Farmer's cheese spread recipe)*
- *1 cup of raisins*
- *2 spring-form pans*

To make the dough:
Using only your hand, mix the flour, butter, 1/3 of the sugar and grated lemon peel. With a spoon, in a separate cup, mix together the sour cream, 1 egg yolk and baking powder. Knead into the dough. Form into a ball, wrap with plastic-wrap and refrigerate for one half-hour.

Use both hands to spread flat it on an un-greased baking sheet.

To make the filling:
Using the whisk attachment on your mixer, whisk together the 3 egg yolks and 1/3 of the sugar, until foamy. By hand, add the farmer's cheese and raisins.

To make the meringue topping:
Using the whisk attachment on your mixer, whisk together the 3 egg whites and 1/3 of the sugar, until hard peaks form. Mix in the juice of one lemon.

To finish it:
Preheat the oven to 350°F. If available, use convection setting.

Separate the dough into the two spring-form pans. The dough is sticky, so cover with the plastic-wrap and flatten with your hand into each pan. Prick with a fork in several places. Bake for 15 minutes.

Spread a thin layer of jam on the dough. If the jam is too thick, mix it with little amount of hot water first. Next, spread the farmer's cheese mix on top of the jam layer. Return to the oven for an additional 30 minutes.

Remove from the oven, and carefully top with the meringue. Place back in the oven and bake

SECRETS REVEALED

for another 15 minutes, or until lightly browned. Remove from oven and let cool.

Cut, using a sharp wet knife, and serve at room temperature.

Appendix A: Dictionary of terms

GLOSSARY
FORD TUNING

Dictionary of automotive terms

In this book, the Index and Glossary are merged into one chapter. Any word in the book in *italics* is defined in this chapter.

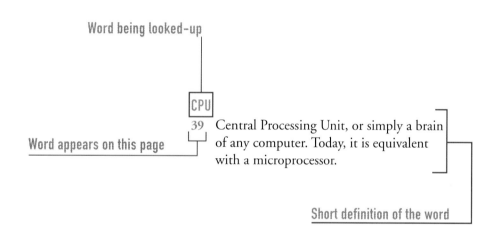

GLOSSARY
FORD TUNING

Address
43 A location of data byte within the memory.

Address bus
42 Connects one chip address lines to another.

BARO
32 Barometric pressure sensor. It measures only barometric pressure. At higher altitudes the air contains less oxygen and the engine requires less fuel.

Barometric pressure sensor
   (see BARO)

Bit
41 Smallest piece of digital information. It is 1/8 of a byte. Bit value can be either a logical 0 or logical 1.

BS0
44 Bank Select 0. One of two signals in Ford Multiplexed Bus introduced with EEC-V. In conjunction with BS1 it selects which quarter (or bank) of the ROM will be accessed.

BS1
44 (see BS0)

Byte
42 Standard data size unit. 1 byte equals 8 bits. Byte value ranges from 0 to 255. An example of a byte is one character. The English word "Hello" consist of 5 characters; therefore its size is 5 bytes.

Bus
41 Another name for multiple signal lines dedicated to the same task linking one chip to another.

Bus width
41 The count of lines within a bus, dedicated to the same task. Typical bus width is 8, 16 or 32.

CARB
118 The California Air Resources Board (CARB) – A board whose mission is to promote and protect public health, welfare and ecological resources through the effective and efficient reduction of air pollutants while recognizing and considering the effects on the economy of the state.

California Air Resources Board
118 (see CARB)

COM port
52 Another name for the serial port on a PC. Since a PC can have more than just one serial port, they are numbered COM1 – COM4.

Compiling
64 A process of translating computer programs from one computer language to another.

CPU
39 Central Processing Unit, or simply a brain of any computer. Today, it is equivalent with a microprocessor.

Data
42 Values or characters operated on by a computer. An example of data is a file.

Data bus
42 Connects one chip data lines to another.

Data byte
42 (see byte)

Data tables
Collection of data bytes, organized into spreadsheet for easier editing.

DIP
46 Dual In-line Package ñ a particular chip shape (type of chip packaging).

DIR
44 Output signal of EEC microprocessor and a control line of Ford Multiplexed Bus. Indicates the Direction of information flow on the bus. DIR is an equivalent to R/W, or Read/Write on non-Ford microprocessor.

Data-bank performance module
98 Performance module containing only a part of EEC program, usually only 1-bank, called data-bank.

DiabloChip
103 "I'm two horny" performance module, manufactured by DiabloSport. Check it out at www.diablosport.com.

GLOSSARY
FORD TUNING

Differential bus
107 A two wire serial network, with "Bus+" and "Bus-" signals, where "Bus+" voltage level is exactly opposite of "Bus-" voltage level at any time. It seems like an unnecessary duplication, since they both carry the same information, but it makes a very good noise resistant system.

DIP
Dual in-line package, an older way of manufacturing Integrated Circuits (chips)

Dynamometer
114 A dedicated machine to measure engine or car performance.

Dyno
114 (see dynamometer)

Dynometer
114 (see dynamometer)

Disassembler
71 Reverse-engineering tool, designed to convert binary machine code back to lost or missing assembly language. It is an opposite task to compiling, where a higher level language gets converted into a machine code.

EEC
31 Ford Electronic Engine Control. Other nomenclatures for the same unit are PCM (Powertrain Control Module – it usually controls transmission as well), ECM (Electronic Control Module) or ECU (Electronic Control Unit).

Emulator
88 (see ROM emulator)

Environmental Protection Agency
118 (see EPA)

EOBD
European On Board Diagnostic standard based on the American OBDII. All vehicles with gasoline engines sold in Europe after January 2001 must be compatible. Compliance of diesel vehicles will be enforced from January 2003. (see also Hellion diagnostic tool)

EPA
118 Environmental Protection Agency – Federal agency. Office of Mobile Sources is the branch concerned with auto emissions.

EPROM memory
Erasable Programmable Read Only Memory is an integrated circuit. Typically packaged in ceramic instead of plastic enclosure, and with a glass

EPROM memory (continued)

window above the chip within. The memory content must be erased with a strong UV light (takes a few minutes) before it can be reprogrammed. (see also ROM)

E2PROM memory

(see EEPROM)

EEPROM memory

Electrically Erasable Programmable Read Only Memory is an integrated circuit. The memory content can be erased and reprogrammed while in use, although relatively slow (takes few miliseconds). (see also EPROM)

Firmware

52 A software written for small microprocessor based systems (like cellular phones or microwave ovens)

Flash memory

47 A modern replacement of EPROM memory. Used the same way, but it can be erased during operation and in-circuit, allowing reprogramming of EEC-V Ford control unit through OBDII diagnostic connector. (see also EPROM; and Predator)

Flowchart

64 A flowchart is a to-do list for microprocessor. It describes any procedure in a most general way in a form of block diagram.

Ford Multiplexed Bus

43 Custom designed port on back of every Ford control unit. After proper cleaning it can be used to attach performance module or emulator. (see also Performance module; and Emulator)

Hellion diagnostic tool

110 Ergonomically designed handheld tool to diagnose any automobile manufactured after January 1996. It is fully OBDII and EOBD compliant. For up to date information visit www.eobd.com.

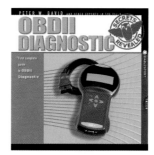

Hexadecimal

55   A numbering system based on 16 digits used on the computers. It uses digits 0 through 9, and letters A though F. Humans use a decimal system based only on 10 digits, 0 through 9.

High level programming language

66   A special kind of simplified English, used by programmers to write computer programs. Later, it gets compiled (converted) toza microprocessor compatible machine code.

Human

40   You must be kidding looking this one up!

IC

39   (see Integrated Circuit)

IHRA

International Hotrod Racing Association

Integrated Circuit

39   Better known as a ìchipî, it is manufactured the same way as a simple transistor, but contains many thousands of them. Integrated Circuit can have many pins (even many hundred) and is delivered in various sizes and packages.

INSTR

44   Output signal of EEC microprocessor and a control line of Ford Multiplexed Bus. Indicates that Instruction (and not data) is currently present on the bus.

International Organization for Standardization

122   (see ISO)

ISO

122   International Organization for Standardization - A non-governmental organization established in 1947. The mission of ISO is to promote the development of standardization and related activities in the world with a view to facilitating the international exchange of goods and services, and to developing cooperation in the spheres of intellectual, scientific, technological and economic activity. ISO is a worldwide federation of National Standards bodies from 140 countries.

J1850 PWM

107   One of the four possible automotive diagnostic formats specified in OBDII standard, which is required after January of 1996. J1850 PWM is generally used by all Ford automobiles. PWM stands for Pulse Width Modulation and OBD stands for On Board Diagnostic.

J3 connector

48   (see J3 test port)

J3 test port
47   An interface to the Ford Multiplexed Bus, accessible through an opening on the back of the Ford EEC unit. Used to connect the performance module or ROM emulator for real-time tuning.

KRAM
41   Keeper Random Access Memory is an integrated circuit. It is an ordinary RAM memory, except it is powered even with ignition key turned off. (see also RAM)

Knock sensor
31   A knock sensor outputs a voltage when it detects engine knock. The EEC then slightly retards the timing to keep it at the border of ideal value.

Lambda sensor
32   (see oxygen sensor)

Live tuning
89   Real-time tuning with a ROM emulator plugged into the J3 test port of the Ford EEC.

Lookup tables
70   (see data tables)

LPT port
52   Another name for the parallel port on a PC. Since a PC can have more than just one parallel port, they are numbered LPT1 – LPT4.

Machine code
67   A microprocessor compatible program ready to be executed. This is what is contained in a performance module.

Manifold absolute pressure sensor
32   (see MAP)

MAP
32   A manifold absolute pressure sensor. It normally measures the manifold vacuum. When the throttle is wide open it also measures barometric pressure.

MB
43   Multiplexed Bus. In Ford EEC it is 8 bit wide bus (MB7..0), and is written-to multiple times to assembly the entire message – hence the name multiplexed.

MFI
    Multiport Fuel Injection. Engine with individual fuel injectors for each cylinder.

GLOSSARY

Memory
42   A chip (or chips) in a computer used to store information.

Memory chip
42   (see Memory)

Microprocessor
39   (see CPU)

Multiplexed bus
43   A version of a bus using multiple signal lines for multiple tasks; linking chips together. (see also Bus)

Multiport Fuel Injection
(see MFI)

Non-volatile
40   A memory type, which hold the information even after the power was turned off.

OBDII
118   On Board Diagnostic standard, version 2. Diagnostic standard, created to lower the exhaust emission and enforced in every automobile sold in USA since January 1996, allowing a single diagnostic tool to be used with any vehicle. (see also Hellion diagnostic tool)

OBDII diagnostic standard
118   (see OBDII)

Oxygen sensor
32   Sensor in the exhaust, measuring amount of the oxygen in the exhaust gasses. It's output values in range of $0\ldots1[V]$, are continuously measured by EEC, in order to achieve ideal air-fuel mixture ratio.

Performance chip
96   (see performance module)

Performance module
96   A circuit board with permanent memory containing a tuned program for the Ford EEC.

PLCC
47   Plastic Lead Chip Carrier – a particular chip shape (type of chip packaging).

Predator
110   Handheld performance and scan tool manufactured by DiabloSport.

Program
40   A procedure (task) converted to computer language.

PROM memory
40 Programmable Read Only Memory is an integrated circuit. Once programmed, the memory content cannot be erased anymore. Also called One-time Programmable Memory. Very often called simply ROM. (see also ROM)

Pulse Width Modulation
123 (see PWM)

PWM
123 Pulse Width Modulation. A way to encode digital information (which only consists of two different values – logical 0 and logical 1) by two different pulse duration. It could be specified that logical 0 transmits as a short pulse and a logical 1 as a long pulse, or vice-versa. A Morse code would be similar to PWM, except it encodes the whole alphabet. Ford uses PWM for diagnostic communication through the OBDII connector.

RAM memory
40 Random Access Memory is an integrated circuit. Information can be both read and written during operation. The memory content is lost without power (this property is called volatile memory).

Reflashing the control module
Reprogramming, or updating the control module, usually through the diagnostic plug using a handheld electronic tool. (see also PROM)

Revolution software
76 Windows based software, causing a real tuning revolution! User friendly graphical charts in plain English can be altered with a click of a mouse. In conjunction with a ROM emulator one's mother can tune Ford automobile.

ROM Emulator
40 Engineering tool designed to perform real-time tuning. In conjunction with a PC one can alter the ROM data contained in the emulator while running the engine.

ROM memory

40 Read Only Memory is an integrated circuit. A family of ROM includes ROM, PROM, EPROM and EEPROM (see each separately). The memory content of ROM is kept even without power (this property is called non-volatile memory). True ROM can be "programmed" only at the manufacturer. The information is etched into silicon and cannot be altered.

Romulator

88 (see ROM emulator)

SAE

122 Society of Automotive Engineers – A 98 year old society which represents the collective wisdom of more than 83,000 engineers, technical professionals, academics, and governmental representatives in 97 countries around the globe. This society is involved in patent protection; solving common technical design problems; development of engineering standards; and the development of a technical knowledge base for industries from the Automotive field to Wet Frictions Systems.

Scanner

A generic name for a handheld diagnostic tool. Some scanners only read diagnostic trouble code. The more useful tool will also read vehicleís live data (such as temperature of intake air; rpm; etc) and freeze frame data (exact snap-shot of circumferences of first memorized trouble code).

Sequential fuel injection

(see SFI)

SFI

Sequential fuel injection. Each fuel injector fires individually in firing order.

Society of Automotive Engineers

122 (see SAE)

STROBE (or more properly /STROBE)

44 Output signal of EEC microprocessor and a control line of Ford Multiplexed Bus. Strobe on this line indicates that data was latched from the bus into the microprocessor.

TBI

34 Throttle Body Injection. An engine with a single or dual fuel injector in a throttle body

Throttle Body Injection

34 (see TBI)

SECRETS REVEALED

Trace
81 A special reverse-engineering capability of a ROM emulator, monitoring an activity of EEC microprocessor and displaying it real time on a PC display. With a little training, it helps to locate the data tables to be tuned.

Transistor
39 Transistor is a basic building block of any electronic circuit. It has three pins – base, emitor and collector.

USB
52 Universal Serial Bus is a relatively new standard for PC trying to replace old and ugly serial, parallel, monitor, mouse and keyboard connectors with a single type and space saving new one. Even though a very good idea, it might not prevail after all. Currently all PCs are for backward compatibility manufactured with combination of ALL of the above connections. Some voices warn USB is not fast enough and software development is too costly.

GLOSSARY
FORD TUNING

NOTES

NOTES

GLOSSARY
FORD TUNING

NOTES

NOTES

GLOSSARY
FORD TUNING

NOTES

NOTES

GLOSSARY
FORD TUNING

NOTES

NOTES

the secrets have been revealed